GEEK DAD

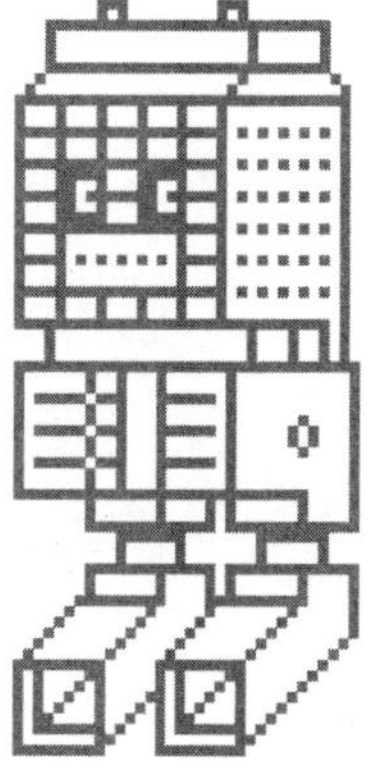

GEEK DAD

Awesomely Geeky Projects and Activities for Dads and Kids to Share

KEN DENMEAD

Foreword by Chris Anderson

Illustrations by Bradley L. Hill

GOTHAM BOOKS

GOTHAM BOOKS
Published by Penguin Group (USA) Inc.
375 Hudson Street, New York, New York 10014, U.S.A.
Penguin Group (Canada), 90 Eglinton Avenue East, Suite 700, Toronto, Ontario M4P 2Y3, Canada (a division of Pearson Penguin Canada Inc.); Penguin Books Ltd, 80 Strand, London WC2R 0RL, England; Penguin Ireland, 25 St Stephen's Green, Dublin 2, Ireland (a division of Penguin Books Ltd); Penguin Group (Australia), 250 Camberwell Road, Camberwell, Victoria 3124, Australia (a division of Pearson Australia Group Pty Ltd); Penguin Books India Pvt Ltd, 11 Community Centre, Panchsheel Park, New Delhi—110 017, India; Penguin Group (NZ), 67 Apollo Drive, Rosedale, North Shore 0632, New Zealand (a division of Pearson New Zealand Ltd); Penguin Books (South Africa) (Pty) Ltd, 24 Sturdee Avenue, Rosebank, Johannesburg 2196, South Africa

Penguin Books Ltd, Registered Offices: 80 Strand, London WC2R 0RL, England

Published by Gotham Books, a member of Penguin Group (USA) Inc.

First printing, May 2010
10 9

LIBRARY OF CONGRESS CATALOGING-IN-PUBLICATION DATA
has been applied for

ISBN 978-1-592-40552-7

Printed in the United States of America
Set in Apollo MT with Armada Light and Lucida Sans
Designed by Sabrina Bowers

To my amazing wife, Robin, who has seen fit to encourage and enable my geeky traits while helping make me the best father I could ever hope to be. This book would not exist without your love and partnership, and I would not be the happy GeekDad I am. I love you!

To my boys, Eli and Quinn, who are enough like me that I can share much of the stuff I geek-out over with you, but are different enough from me that you show me new things every day. You are the reasons I did this. Grow up strong and geeky!

And to my parents, Ellen and Walter: I don't imagine the path I've traveled has been quite the one you expected, but I figure you're pretty happy with where I've ended up. Thank you so much for starting me out on that path with such good preparation and support, and for being there for me every step of the way.

Contents

AWESOME ACCESSORIES

GEEKY KIDS GO GREEN

BUILD/LEARN/GEEK

GEEKY POTPOURRI

Special Thanks

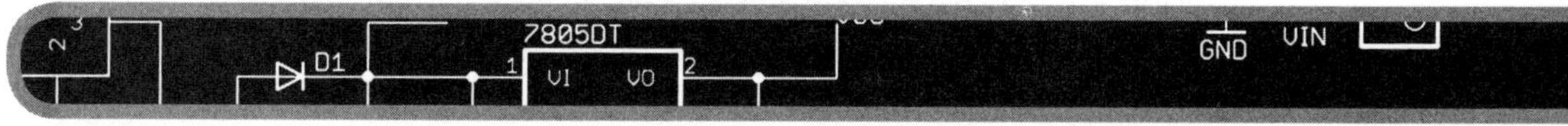

Three years ago, a man, of whose fame and shrewd intelligence I was only then vaguely aware, put out a call for volunteers to write for a blog called GeekDad. I took what I thought was an outside shot, and was lucky enough to be accepted by Chris Anderson to contribute.

Six months later, he asked me to take over running the blog, and once I'd scraped my jaw off the floor, I enthusiastically accepted.

The time since then has been an adventure, and a tectonic shift in the direction of my life. I am living a life I could barely have imagined and have been lucky enough to pay the favor forward to other geeky parents who have come to write for GeekDad, too. But it all started with the entrepreneurial generosity of Chris Anderson, the founder of GeekDad, to whom I am eternally grateful.

A whole new world

Of critical import in the whole "I got to write a book" process is being given the chance. Very special thanks go to Megan Thompson at LJK Literary for "discovering" me and nursing me through the proposal process, and Jud Laghi for getting my proposal looked at by all the right people.

And "all the right people" would be my editor, Lucia Watson, who helped me take a bucketful of cool ideas and present it in the (hopefully) fun and readable tome you now hold, and assistant editor Miriam Rich who has guided me through the strange new world of publishing.

I realize I've been very lucky to have fallen in with such patient and professional people, and I can't express my gratitude for how much they've done for me deeply enough. Thanks!

And there's no way I'm leaving them out

The real success of the GeekDad blog comes from its family of writers, and I can't take credit for this book without giving some back to the team that helped make it all work: Anton Olsen, Brad Moon, Chuck Lawton, Corrina Lawson, Curtis Silver, Daniel Donahoo, Dave Banks, Don Shump, Doug Cornelius, Jason B. Jones, Jenny Williams, John Baichtal, John Booth, Jonathan Liu, Kathy Ceceri, Lonnie Morgan, Matt Blum, Michael Harrison, Moses Milazzo, Natania Barron, Paul Govan, Russ Neumeier, Todd Dailey, Vincent Janoski, and the Mystical Magical "Z."

Additional thanks go to Matt Blum, my right hand at running the blog, for his help copyediting the manuscript, and Bill Moore, Dave Banks, Russ Neumeier, Andrew Kardon, Brian Little, and Natania Barron for contributing projects to this book.

And thanks to the crew of Starbase Phoenix: On the edges of known space, a fire burns to light the way. You guys are that fire.

Foreword

by Chris Anderson, Editor in Chief of *Wired*

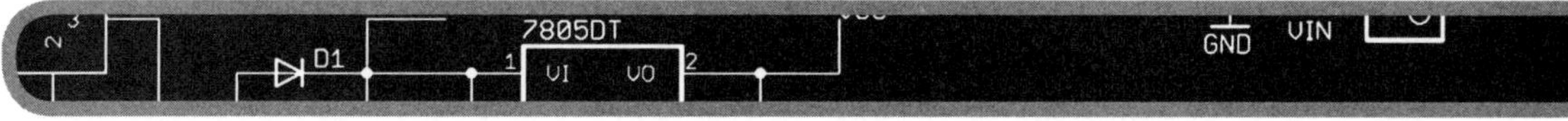

Here's the challenge of being a GeekDad. You're a geek. You're also a dad. Geeks want to do cool projects, ideally involving science, technology, and anything that comes from Japan. Dads, meanwhile, want to spend time with their kids, ideally doing something kids want to do. Most of the time, these two forces are in opposition. But they don't have to be!

The origins of this book, and the Web site that inspired it, were in finding ways to reconcile the call of the geek with the nature of the parent. In early 2007, I started GeekDad mostly for myself: I had four (now five) kids, all under ten at the time, and just could not bear the thought of playing Candyland one more time.

I was looking for projects and activities that were both fun for them and fun for me. Not fun for me and boring for them (most of my geeky stuff) or fun for them but boring for me (most kid stuff), but fun for both of us. In other words, a worthy challenge for all ages.

I hoped that there were other people out there with the same ambition who might respond to my posts. There were. Today, a couple years after the site launched, it attracts more than a million readers per month and has more than two dozen contributors. Led by Ken Denmead, who has run the site since late 2007, GeekDad.com, now an official *Wired* blog, has become one of the top parenting sites on the Web. It turns out that the nexus of geekdom and parenting is a rich seam indeed. Now Ken has taken it to the ultimate degree: the book you're reading. If only such a thing had existed three years ago when I needed it!

My own quest for the perfect Geek/Dad intersection began with LEGO Mindstorms robotics. About three years ago, my boys (then nine and six) were hugely into LEGO, while I was getting into robotics. I'd also been given an RC plane, which the boys and I had tried relatively unsuccessfully to fly. We'd also been given a LEGO Mindstorms NXT kit and had dutifully made all the robots in the instruction manual and left thinking "what next?"

It was clear we were never going to be incredibly good at either LEGO Mindstorms or RC planes, given all the amazing things people could do with both, as evidenced by the videos we looked up on YouTube. And frankly, as a geek, I couldn't see the point in doing stuff that other people had already done a lot better than we could.

But while out on a run, I got an idea. The sensors available for Mindstorms were pretty cool, including gyros, an electronic compass, and accelerometers. NXT also had a Bluetooth and was compatible with other Bluetooth devices, possibly including Bluetooth GPS modules.

What do you get when you put that all together—gyros, accelerometers, GPS, and a computer? An autopilot! If we couldn't fly the RC planes well, maybe we could invent a robot that could. And for all the cool things that people had done with Mindstorms, the one thing nobody had done yet was to make it fly. A worthy project had arrived! We would design the world's first LEGO Unmanned Aerial Vehicle (UAV)—a fully autonomous LEGO-piloted drone.

We started with the mechanical bits. At the time, there was no way to drive RC servos directly from Mindstorms, so we designed a sliding tray with a Mindstorms motor that slid it back and forth, and mounted the plane's rudder servo on that. This way you'd always have RC control, but the Mindstorms controller could override it by sliding the whole servo forward and back. Likewise for turning the autopilot on and off. In the absence of an electronic link between the RC and Mindstorms worlds, we settled for a mechanical one: a servo strapped to a Mindstorms touch sensor.

Then it was time for the software. My son and I had worked on the code all weekend. It had turned out to be a perfect father-kid project: fun for him and fun for me—a meeting of child and grown-up interests turned into a fantastic weekend activity. We started with the simplest autopilot, which would use the compass sensor and just fly in a given direction for a preprogrammed length of time, then switch to another direction for another duration. That's just the navigation part of an autopilot, however; the stabilization part would have to be covered by a commercial stabilization unit called the FMA Co-Pilot.

One flight revealed that it actually worked (more or less)—the plane navigated on its own and stayed in level flight throughout. We posted online about the project and got a huge amount of interest; the idea of a LEGO UAV was just as mind-blowing as we'd hoped. But real aerial autonomy goes a lot further than we had taken it so far: It should include latitude and longitude waypoints, and a proper autopilot should handle both navigation and stabilization. On to version 2.

This one was much more advanced. By now, thankfully, HiTechnic, a company that develops Mindstorms sensors, had gotten in touch and offered us a prototype of a device that not only allowed Mindstorms to drive RC servos directly, but also handle the autopilot on/off switching from the RC system. Others, led by Steve Hassenplug, an amateur Mindstorms guru, had figured out how to interface Bluetooth GPS modules with Mindstorms, so it was now able to use standard latitude and longitude waypoints. And HiTechnic had also made a prototype "integrating gyro" that simplified the math of working with such inertial sensors.

Combining all of those with HiTechnic's three-axis accelerometer made the basics of a real Inertial Measurement Unit, the core of the most sophisticated autopilots in the world. But with a LEGO

Mindstorms controller at the core! We mounted it in a plane, did a heap of programming, and more or less proved that it could work.

The LEGO UAV 2 flew a few times to prove the concept, then retired to a career of trade shows and the LEGO Museum in Billund, Denmark, where it is today. The kids, sadly, left the project before it was finished. By the time the code had switched to Robot C and the soldering iron had come out, we were well beyond any conceivable Geek/Dad balance. But the seed had been planted—and this was fertile soil.

The search for similar geeky projects that cut across generational appeal led me to start GeekDad. My jokey early motto was "Permission to play with cool toys isn't the only reason to have kids, but it's up there." The point was to focus on the Venn intersection of geeky interests and parenting: to find cool science/tech/culture things that are fun for all generations.

I bought the domain for not much from a nice guy who wasn't using it, and then started blogging intermittently. Then I invited friends to join me, and soon put out an open call for other geeky dads to participate.

The rest, as they say, is history. This book is what I could only dream of three years ago. My kids and I will happily dip in on weekends, picking projects that fit the time and materials we have at hand. You should do the same. It's not meant to be read front to back. It's a book of ideas and instructions. Skim, share with your kids, and find something that sounds like just the thing to fill a Saturday afternoon.

When you're done, you will have had an adventure, that's sure. But you may have also triggered a curiosity in your child that could lead to a lifetime fascination. You had such a moment when you were a kid; that's what made you a geek at heart. This book is about opportunities to create such a moment. And to be a great parent while doing it.

Have fun!

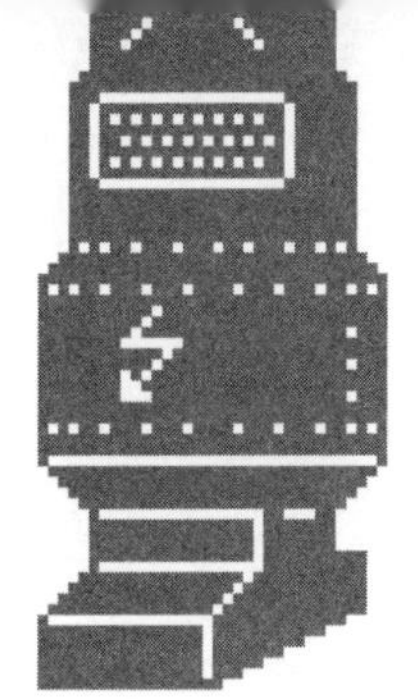

GEEK DAD

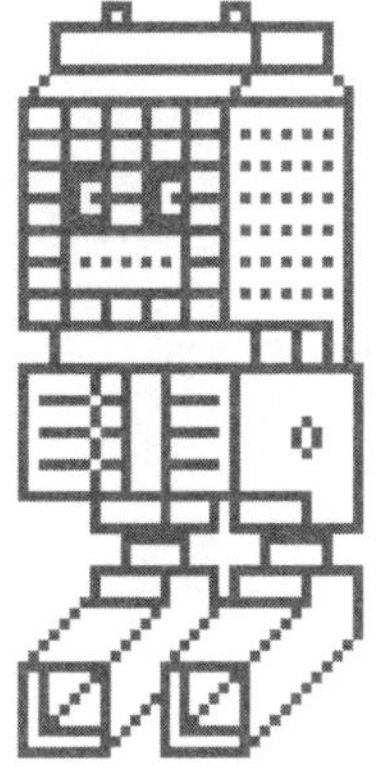

Introduction

About Being a Geek and a Dad

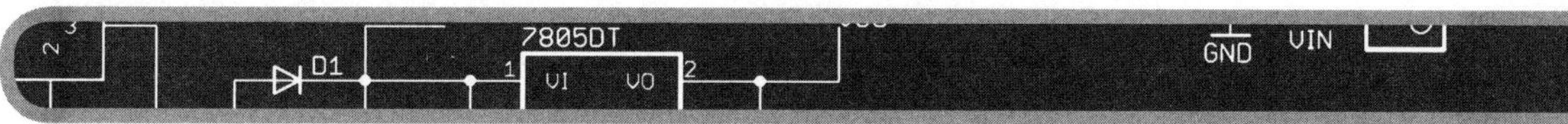

Once upon a time, the word *geek* was used to describe circus performers. Then it evolved as a pejorative to describe awkward, skinny kids who got routinely thrown into school lockers by the high school football team. But these days, *geek* has reinvented itself. This is the era of the geek. And geeks are cool.

There is some interchangeability between *geek* and *nerd*. They both generally describe someone of restricted social ability who finds enjoyment in pursuits outside the mainstream—pursuits like computers, role-playing games (RPGs), science fiction and fantasy literature and movies, science and engineering, and so on—you get the idea. But there is a key difference between the geek and the nerd.

One renowned geek dad (and honorary GeekDad), Wil Wheaton, describes it pretty simply: A geek is a self-aware nerd. It makes a lot of sense to me—I think geeks had those social issues growing up and liked all those things that weren't part of the popular culture in school, but we came to understand our nature and, in a very Kübler-Ross kind of way, moved past the self-limiting aspects of nerdhood to a state of acceptance, and even enjoyment, of our place in the universe. Which, in a funny way, helped us take care of some of those social issues, because a lot of us ended up actually getting married and having kids (which totally rocks!).

I think part of the current ascendancy of geeks in general, and GeekDads specifically, is that there are a lot more geeky women than people realize, and some of us geeky guys were smart enough to recognize our own kind and attempt to mate and perpetuate the subspecies.

But before I get too far along, let me point out something important: Geeks aren't just about the computers and the D&D and the passion for anime and comic books. There's a whole lot more out there that people get passionate about, even mildly obsessive about, that can qualify them as geeks. If you're so passionate about something that you're not just good at it but can lose yourself doing it for long periods of time (often to your social detriment), you may be a geek. If you carry encyclopedic knowledge about a topic and will joyfully use it to act as the pedant whenever the subject is being discussed, you may be a geek. If you have a room in your house devoted to a hobby that other family members avoid talking about, you may indeed be a geek. I'm not talking about "experts" or "professionals"—I'm talking about the real deal. Here are some examples:

SUBJECT	FAMOUS PRACTITIONER	GEEK
Guitar Player	Eric Clapton	Buckethead
Chef	Emeril Legasse	Alton Brown
Film Director	Martin Scorsese	Peter Jackson
Wide Receiver	Steve Largent	Jerry Rice
Playwright	Tennessee Williams	William Shakespeare
Video Game Designer	Anyone on a Madden Title	Will Wright

So, what are the factors that make up the geek? I'd like to posit that the geek is a combination of common personality factors that we see in all sorts of people. Indeed, these factors taken alone or only in pairs may lead to less desirable characters. See, for example, the Venn diagram below (talk about geeky!), where I've described the possible combinations of key personality factors that make up the geek, and its associated stereotypes: Knowledgeability, Obsessiveness, and Social Skills.

Knowledgeability represents having significant stored information with easy recall. That knowledge may be broad and relatively shallow—the know-it-all—or it may cover only a few topics but be deep and profound—the expert/problem solver.

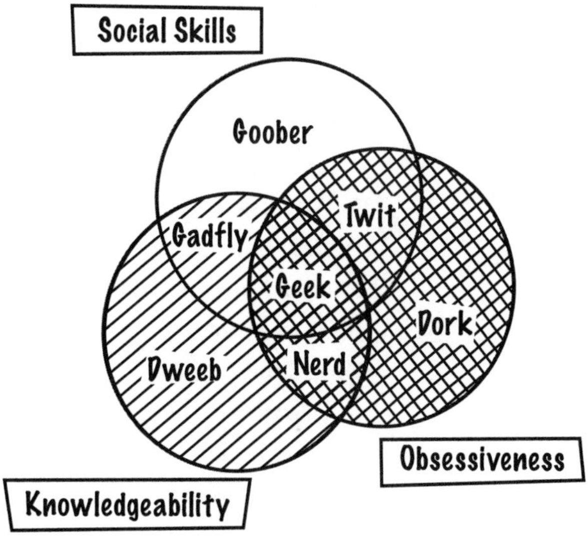

Obsessiveness is a person's ability to lose himself in something he has a passion for. Common symptoms include losing track of time while coding HTML/CSS or staying up until four A.M. to finish Portal because you had to earn watching the final credits (and hearing that awesome Jonathan Coulton song).

Social Skills can mean a lot of things, not all of which are about being "popular," which geeks and nerds always feel they never were in their formative years. But geeks do at least have enough presence and personality to form lasting relationships, which helps differentiate them.

So first, it's easy to tag all the stand-alones: Dorks are the people who are obsessive without the introspection to recognize it in themselves or how it could affect others. Dweebs know everything but can't apply or express themselves. Goobers are good-natured but lazy idiots—no one minds them, but they aren't much use.

It starts to get interesting when you begin combining the traits.

The classic nerd has knowledge/intelligence AND the obsessive nature that produces results. You can't expect them to carry on conversations that won't lose a non-nerd audience—they would talk your ear off about something as nerdy as the exciting application of quantum theory on the flow of mold over a piece of cheese, but set them to work on a project without distraction, and you'll be able to mine the results for pure gold (especially if it has to do with World of Warcraft and, you know, gold mining).

The twit—well, I suppose there are other names for this person, probably a lot of regional variations—but the twit combines obsessiveness and social skills into a double-edged sword. This could be that sales guy who can talk up a storm but who really doesn't know squat, or it could be the diligent hard worker everyone likes but who really just doesn't get it.

And then there's the gadfly. He's smart and he gets invited to parties, but he's lazy. Or worse, he's intellectually smart but emotionally ignorant, and doesn't care. He's the one most likely to be the pedant in any gathering, and he probably uses people to get the work done he finds beneath him.

Of course those are extremes, and there are perfectly lovely, functional people who fall into those categories; but they're not the ones we're here to talk about. In the sweet spot, right there in the middle, is the tripartite synergy that creates the geek. The mixture of knowledge (about comic books, particle physics, or the works of Mozart), obsessiveness (they'll sit in front of a computer or a workbench for hours perfecting, building, or playing anything), and social skills (they actually get together with people for pen-and-paper RPGs or get in line with a bunch of friends to see the midnight showing of the next Star Trek movie), that makes a well-rounded, self-sustaining person of affable oddity.

Now maybe weigh it just slightly toward the social skill set, and you have someone who can actually get a date, find a mate, get married, and procreate. That, in a nutshell, is how a GeekDad comes into being. The conditions need to continue to be favorable—is there support at home for ongoing geekiness? Will infecting the child(ren) be allowed? How many times will the wife feign a chuckle when you lift your little tyke and in a deep voice intone, "Luke, I am your father" (knowing it's a misquote) before it gets old? How many jokes about containment breaches will be tolerated at diaper-changing time?

It helps immeasurably when your mate is a geek, too (but that's another book). I've been lucky enough to have that situation in my marriage. In fact, not only have my little quirks been tolerated, but some of them have actually been encouraged. And in return, I encourage back. I mean, how many men can say their wives wanted a trip to a science fiction convention for their anniversary? I'm one lucky man.

But the best part is getting to share with my kids, share the geeky things that informed my childhood and continue to inform my existence: Star Wars, Star Trek, math, science, reading, writing, music, computers and video games, movies and television. I can't tell you the joy of having my kids get into *Doctor Who* and comic books and *Lord of the Rings,* and then talking with them about the important aspects of the stories and watching them just soak it up.

I lived through the school years as a breed apart (though I had good friends who were geeks, too), so it makes me feel great to be able to inform and guide my kids through the social aspects, and the occasional challenges, of growing up as a geek. All parents want to protect their kids, but I like to think the best protection I can offer them is to help them understand what will happen, why, and how to best deal with it. I want them to know that different isn't bad, and that being intelligent and inquisitive is something to be proud of.

Indeed, that's what being a GeekDad really means for me. For all our personality quirks and interests in pursuits that are outside the mainstream (or at least interests more technical than is usually palatable for the mainstream), we're all about understanding, and communicating, and connecting with others by sharing what we love and helping others to grok it as well. Of course there's a biological imperative to have kids and raise them to survive and thrive, but we want them to be happy, too—whatever happiness may mean to them.

I'll encourage my kids to love what I do, but I won't force it on them, and when they want to try something different, I'm happy to let them just as long as they come at it like a geek: They should be knowledgeable about it, be a little obsessive about it, and get along with the other people who are doing it. That's what all the greatest geeks do.

GEEKY PROJECTS FOR DADS AND KIDS TO SHARE

Most "parenting" books aren't about things you can do with your kids. Most are about things to do *to* your kids, tricks and tactics for tweaking their behavior in some desired manner usually at odds with what kids really want: to play, and spend real quality time with you.

I'm not saying all those books are bad. Some of them do try to reinforce the idea of spending quality time (though I'd really like to find a new phrase to replace *quality time*) with your kids. This book has the same goal of those others: to help you share time with your kids in their formative years in constructive, educational ways, without making that time seem as if it's supposed to be constructive or educational (not always easy). The difference here is that from a geek's perspective, constructive and educational may not mean what all those other books think it means. Here's what makes our approach different:

- Geeks like games that require a fantastic imagination.
- Geeks love science and knowing how things work. Experi-

mentation is the best way to learn those things. If things go "boom" in the process, all the better.

- Geeks love finding interesting, creative solutions for problems that could be solved in a more mundane fashion.
- Geeks love to play, but in playing, to build and learn as well.

There is a plethora of projects included here about an eclectic array of subjects, from board games to electronics, crafts to coding. But I'm not here to tell you exactly what to do. The instructions are meant give you a structure to start your adventure with your kids. Each of these projects will allow for extensive customization and personalization. Indeed, what I have in my workshop and available at the hardware store in my town may be rather different from what you have. So I expect you to improvise, adapt, and even (quite likely) improve on these projects.

PROJECT INFORMATION

At the start of each project, you'll see a table with summary information to give you an idea what to expect from it, and there are some symbols not unlike what you see in a restaurant or hotel review to explain cost and difficulty. Here's a legend to explain their meaning.

PROJECT	TITLE OF THE PROJECT
CONCEPT	A quick overview of the project so you can decide if it's of interest to you
COST	$ = $0 to $25 $$ = $25 to $50 $$$ = $50 to $100 $$$$ = $100 on up In many cases I'll exclude the cost of tools and materials that are so common that you probably already own them or could find or borrow them easily.
DIFFICULTY	= easy for primary school-age kids to grasp and enjoy = for secondary school-age and up = for junior high and up = high school age There is a wide variety to these projects, from the very simple to the rather complex. Some of them can even be adjusted to be more or less in-depth. It's up to you to gauge your kids' ability level and attention spans, and pick the right projects to share with them. After all, you know your kids better than I!
DURATION	= 0-15 minutes = 15 minutes to 1 hour = 1 to 3 hours = 3 hours and longer While the journey is often just as important as the destination, it's nice to get there, too. You and your child can finish most of these projects in an afternoon or evening of cooperative work.
REUSABILITY	= one time only = re-use once or twice = multiple reuse possible = good forever Hopefully, you'll find that most of the projects in this book will have serious repeat value.
TOOLS & MATERIALS	A list of the basics required to perform the project. I'll do my best to suggest how to make do without buying too much.

One thing you'll notice as you go through the projects in this book is that they are not long, costly, or overly difficult and involved projects that take too much work before paying off in the fun department. If you and your kid have the kind of patience and geeky determination to spend days/weeks/months on a project, then let me suggest you take up painting Warhammer armies or mapping the visible sky in your area with a telescope you built from scratch.

It's not that I don't have respect for folks that do that kind of thing! On the contrary, they are the epitome of geekhood, and I am not worthy to clean their brushes or polish their lenses. I just don't have that kind of time or energy. I want to do something fun with my kids NOW (or at least in the few minutes to couple of hours it takes to complete any project in this book). So you'll find that the most important common features all these projects have is that they are accessible, affordable, and truly buildable for just about anyone with an ounce of geek in them.

Okay, it's time. Go get your kid(s) and get started!

MAKE YOUR OWN GEEKY GAMES AND CRAFTS

Make Your Own Cartoons

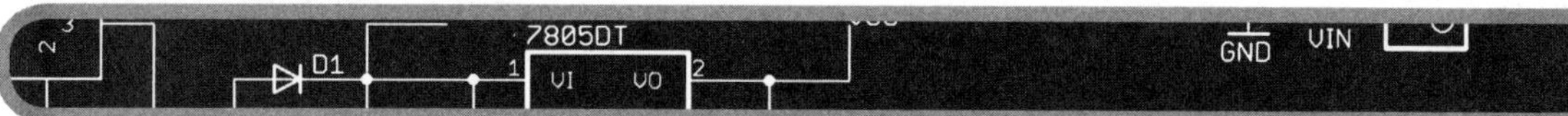

There aren't many folks, geeky or not, who don't love cartoons. Beyond the Sunday paper or *The New Yorker*, there are cartoons out there that have particular appeal for us geeks: Dork Tower, xkcd, or PvP Online. Reading them is fun, but as with most things, wouldn't it be even more fun to try to create our own?

Maybe you or your kids have already had a good idea for a comic story line or comic characters of your own. Don't worry if your drawing or graphics arts skills aren't up to the task—this project will help you overcome that niggling deficiency. Instead of pen on paper, you can use the tools you're familiar—even handy—with to create something visually distinctive, creative, and all your own.

PROJECT	MAKE YOUR OWN CARTOONS
CONCEPT	Make comic strips using digital pictures.
COST	$$–$$$$
DIFFICULTY	
DURATION	–
REUSABILITY	
TOOLS & MATERIALS	Digital camera with decent zoom/macro settings; light box; action figures; computer; software

The basic idea for this project is to use LEGO minifigs or other similar action figures or toys as your cartoon characters, photograph them rather than draw them, and then manipulate those images on your computer to create the comic. Before you actually get to photographing, you need to sort out the basic features of your project.

CHOOSING YOUR SUBJECT

What's your strip going to be about? Is it you and your kids, or some imaginary characters? I decided to base my strip on the amusing Twitter messages of well-known geeky writer/actor/dad Wil Wheaton. Wil has the habit of tweeting the imagined conversations he has with various programs on his computer (especially iTunes) or with his dog. Because of the character-length limitations of Twitter, these conversations often take on the form of quick scripts that I realized would fit perfectly into standard six- or nine-panel comic strips. You may decide to use amusing vignettes from everyday life, outrageous things your kids have said or done, or just about any other story that can be told in three to nine panels.

CREATING THE FIGURES

You need to make your characters by using some kind of action figure you can pose and modify depending upon the story you're telling. To make Wil as a cartoon character for my project, I discovered a really easy idea: I turned to LEGO minifigs, the little people that come in many LEGO sets. Here's how you can create your own characters:

1. Go online to the shopping section of LEGO's Web site (you might have success at your local LEGO store as well) and buy a few stock minifigs (lower and upper body, and head with no special features).

2. Then, in the à la carte area, purchase an extra dozen basic minifig heads (called Mini Head No. 1 at the LEGO store—it's the one with the simple smile).

3. When you have them in hand, you can customize them to match your desired characters. I found an image of Wil online that is well known to his fans (he's wearing a very particular clown-face sweater) and re-created it on the minifig torso, using fine-tipped felt markers. If you have trouble getting a minifig torso without any markings on it, it's easy to scrape the markings off the plastic with an X-Acto knife or razor blade.

4. To give you a palette of emotions for your cartoon character, use the extra minifig heads. Just like the bodies, the printings on the heads are easily scraped off with a blade. For my Wil minifig, I took one head, scraped off the generic smile, and drew in a surprised "oh" look with a fine-tipped black Sharpie. If your script has your character going through many adventures and scenes, you might want to create sad faces, happy faces, startled faces, and so on. Since the heads are interchangeable on the body, you can have as many different looks as you can think of for the same character.

Your kids can have fun crafting their own characters. You can also use minifigs for stock characters in your script. There are all kinds of minifigs available, so coming up with an astronaut or cowboy or law-enforcement officer is pretty easy.

WRITING THE SCRIPT

It's always a good idea to start a comic strip project by writing a quick script to break down each shot so you know exactly what figures you need and what pictures you'll have to take. Even though the idea of this project is to avoid actual illustration, work with your child to sketch out a visualization of your idea, blocking out the shots and scribbling in the words or thought balloons you'll need. This will help you figure out what kind of layout you want—how many panels, what size, and what relation to each other. And don't forget that sometimes a panel with no words can be the funniest of all!

BUILDING YOUR STUDIO

The setup you use for photographing your figures can be very simple and cheap, or very involved and expensive, depending upon how complex you want to get, and how polished a final product you want to create. Here are some tips:

- The most expensive tool is also the one you're most likely to already own: a digital camera with a zoom lens. Your child may even own one.
- A tripod will be invaluable help. You will need to set up shots that look very similar from frame to frame, with only slight changes to the position of your characters or to their facial expressions. Having the camera locked down in one spot is vital to achieving that.
- You'll want a light box of some kind to properly light your subjects and to create a background as blank as possible, to

simplify the image editing later on. You could start as simply as some white poster board taped into four sides of a cube (bottom, back, and two sides—leaving the top and front open for lighting), and then some inexpensive lights mounted above to give good enough flood lighting to minimize shadows. A very tiny step up would be to purchase one of the portable light studios available online or at most stores that sell cameras. These usually don't cost more than $40 and will do the job admirably.

SHOOTING YOUR SCRIPT

Once you have your equipment set up, it's time to take pictures—and plan on taking a lot of them. You'll probably want to start out with some basic shots to test the lighting and exposure, and, while the auto settings on your camera may work well, you might want to consider using the manual settings to ensure that every picture you take has the same exposure and color balance, for continuity of look. I used an inexpensive DSLR mounted on a tripod, with a fairly slow shutter speed and a delay so I could hit the button and step back before the picture snapped, just to get things really sharp and well lit.

Use the script you worked out earlier to set up your scenes, but don't skimp on taking pictures. Set up each scene a couple of ways, play with the blocking of your characters (where they are in relation to one another), and take a series of pictures with slight differences in stance, perspective, close-up, long shot, and so on. It's always better to have multiple choices for each frame when you get to postproduction, rather than being unhappy with something and having to set up and shoot again.

And don't worry much about the resolution of the images, either. You don't need to be shooting in your camera's RAW format here. A typical 3-megapixel resolution will do just fine, and you can fill your image basket with hundreds of shots to choose from when it comes time to start composing your strips.

ASSEMBLING YOUR STRIP

Now comes the composition stage. If your child is younger, you can run the keyboard and s/he can act as artistic director. If your kid is older or very savvy with the technology, now's your chance to step back and watch his imagination run wild.

You have a range of options available to you for actually building the strip:

- You could use something as simple as Microsoft Word and create the strip frames from Tables by pasting your images into the panels and using text boxes for the lettering.
- You could also build the strips in many of the graphics programs out there—Photoshop Elements, GIMP (GNU Image Manipulation Program), or Pixelmator come to mind.
- There are also programs dedicated to crafting homemade comic strips. I used Comic Life Magiq ($50 from www.plasq.com) for the Mac, because it has great templates and image manipulation tools so you don't have to touch up your pictures before getting them assembled into a comic.

When you've settled on the tool, it's time to get creative. Everything is up to you—what do you find funny or compelling? How do you like seeing a story or a joke told? Here are a few things to think about:

- Since so many comics are almost literally made up of talking heads, assembling your panels may be as simple as using the same picture over a couple of times and just slightly adjusting the orientation so that it looks like a different shot.
- As they do in movies, moving from a long shot to a medium shot to a close-up can help drive home a dramatic or comic statement.
- Never forget that a panel without dialogue is one of the most powerful comic beats. Feeding the punch line in one panel, and then following it up with the identical panel sans words, or even better, having a character react directly to the reader (called breaking the fourth wall), can be comedy gold.

In the end, depending upon the subject matter and the tools you use, your final product may look something like this (I went with an imagined conversation with iTunes here):

The best thing about this technique is how easy it can be to build up a store of images—close-ups of different expressions, or generic interactions between regular characters—to the point where you won't always need to shoot new images when you come up with a new idea. After a while, you'll be able to simply pull from your archive of shots to build new strips. In this way, any time a new idea hits you and your kid, the payoff can be very quick (while teaching a valuable lesson about planning ahead and being prepared for the future). Plus, you know, it's fun.

The Coolest Homemade Coloring Books

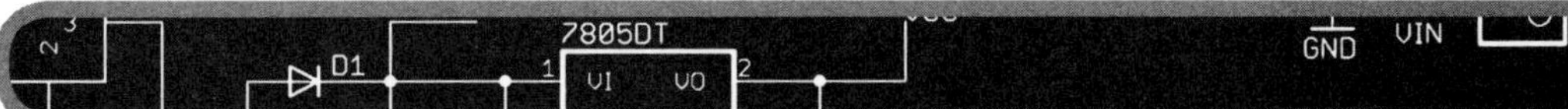

From the moment most children are old enough to pick up a crayon, coloring is a creative pastime they enjoy. And please take special note of that word: *pastime*. How many times, on a rainy day at home perhaps or when you know they are stuck in a waiting room with you, have you been desperate to NOT let your kids be babysat by a video game machine or the television? Coloring can be an absorbing, constructive, imaginative way of passing the time—even for older children. And it can be just as portable as any Nintendo DS or PlayStation Portable (PSP).

But the selection of available coloring books is always either bound by current pop culture or skewed to younger children, and can put older kids off. What if you could make your own coloring books for your kids, filled with images for them to color that come from things they actually love and will get absorbed in? Well, you can, and pretty easily, too. Here's how.

PROJECT	THE COOLEST HOMEMADE COLORING BOOKS
CONCEPT	Use any images you or your kids choose to make dot-to-dot sheets or coloring book pages.
COST	$-$$
DIFFICULTY	⚙⚙
DURATION	☼☼–☼☼☼
REUSABILITY	⊕
TOOLS & MATERIALS	Blank paper, a binder, crayons or pens, a computer, scanner (optional), common graphics software (Photoshop Elements, Pixelmator, GIMP)

Because there are levels to artistic ability and interest, I'll show you two different coloring pages you can make. First, the classic dot-to-dot, then coloring sheets. They are all made from images you can find online or scan into your computer from sources at home.

DOT-TO-DOT COLORING PAGE

To start, you have to pick the image you want to turn into the coloring page. For dot-to-dots, it's going to be a lot easier to select simple images, though if you're patient and want more detail, you can go for the gusto with more detailed pictures. But dot-to-dots are great for younger kids who are learning about staying within the borders. And they love to be amazed by what they can create by drawing a series of lines between dots.

For this example, we're going really simple—an image of the sigil of the Rebel Alliance.

(Advisory: If you are reasonably experienced with graphics software and know what a layer is, you can skip down a couple of paragraphs to the one that starts "Select a pen . . .".)

Assuming you're a geek, we are going to figure you have some manner of graphics software, probably for touching up pictures from your digital camera. Common (and pretty good) examples of such software include Photoshop Elements (the cheaper, easier-to-use version of the industry standard Photoshop) available on Mac or PC; Pixelmator for the Mac; or GIMP, which works on PC, Mac, or Linux and is FREE FREE FREE. Maybe you haven't played around with the software that much. If that's the case, here's a quick lesson about the first feature you're going to use.

A layer is a standard concept in image editing software. Conceptually, it's very simple. Imagine if you took a printed picture and laid a piece of tracing paper over it, and then copied the features by hand on that tracing paper. Well, a layer is just like tracing paper, only it works digitally on the computer, and you can have as many of them as you want.

STEP 1: For this project, you open your base image in the editing software of your choice. Then you add a layer. Most of these pieces of software have a menu actually called "Layer" from which you can "Add a Layer." Once you've done that, there is a perfectly transparent layer of digital tracing paper on top of your image, upon which you can now trace, without affecting the original picture.

STEP 2: Select a pen or pencil tool and a fairly small brush size to make your dots with. Draw black dots all around the edges of the image, at relatively even intervals. Straight lines need only one dot at either end. Curves need more so that, when they are connected, they will better re-create the curve.

STEP 3: These programs should have a separate control window that shows all the layers in the current project. From this window, you can now "turn off" the layer with the original image on it so that you see only the dots.

STEP 4: If you want to go all the way, you can also use the program's text feature to add numbers next to each dot to give your artist a sequence to follow. Or you can just save this file and print as many of them as your kids want to color, letting them be creative and decide how to connect the dots.

COLORING BOOK PAGES

Obviously, older kids and those with more advanced coloring skills are going to be hankering for something a bit more challenging than connect-the-dots. What's great is that these graphics programs have filters that let you kick out coloring pages by the ream with only a couple of clicks. They require even less work than the dot-to-dots.

Pick an image. For this example, I used a snapshot I took of my classic Nauga (www.nauga.com/promoitems_nauga.html) in my office. I opened the image in Pixelmator and then used "Filter-Stylize-Line Overlay" to automatically find the edges in the image and drop everything else out, making a perfect coloring sheet (there are settings you can tweak to get it "just right").

In Photoshop Elements, the process is nearly as easy. Open the image, and use "Filter-Stylize-Find Edges." Then use "Enhance-Convert to Black-and-White" to drop out the colors, and you have much the same effect.

In GIMP, you can try "Colors-Desaturate," then "Filters-Edge Detect-Neon" and "Colors-Invert" to get a similar effect. You may need to play with some settings to get an optimal result (and you can save those settings for future uses). GIMP is just as powerful as the other programs in many ways, but it is not quite as user-friendly, so there's a bit more of a learning curve.

Once you have the technique down, you can whip these out en masse and build your kids (or get your kids to build) their own coloring books, using images they find online (Google Image Search is excellent for this, though make sure you keep an eye out for inappropriate content; or try the Web sites for the cartoon shows they like—Disney or Nickelodeon) or scans from other books or sources.

One other way to do this—with slightly less creativity (and therefore less geek factor) but without the need for special software—is the Coloring Page Maker at the Crayola Crayons Web site: http://play-zone.crayola.com/play-zone/index.htm.

Create the Ultimate Board Game

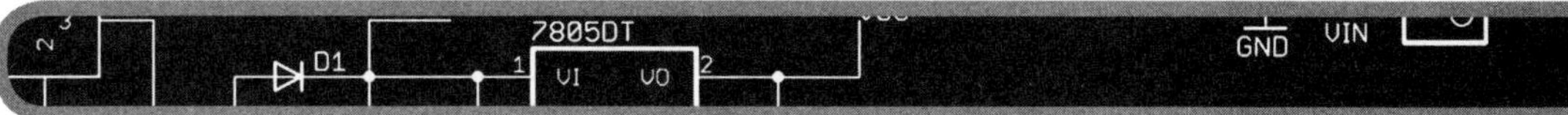

In an age when the video game seems to be king, it's interesting to notice that when you walk into any mega-mart toy section, you still find whole walls devoted to the board game in all its varieties. Everyone still loves the low-tech joy of Life, or Risk, Chutes and Ladders, or other classic board games.

For many (including me), one of the best board games was Mouse Trap (not to be confused with the Broadway play), where you got to build a homemade Rube Goldberg device to catch a plastic mouse. If you didn't play it on a perfectly flat, level table, it was often a challenge to make it work just right, but when it did—magic!

But when you look at any board game, when you strip it down to its core, there are elements that are common to them all: playing pieces that move a number of spaces based upon random number generation; special spaces that do interesting things; special cards that grant a boon, or curse the players; a final space to be reached representing the end of a harrowing journey. In short: adventure. Dress up the structure of the game in whatever outfit you like—the struggle to reach the top of a path (Chutes and Ladders), the simulation of a modern life (Life), the quest for world domination (Risk)—most of the basics are still there.

So, given the hacking, maker spirit of the geek, what's to stop us from making our own board games? Nothing, I say!

PROJECT	CREATE THE ULTIMATE BOARD GAME
CONCEPT	Build and customize your very own board game based on this simple concept.
COST	$–$$
DIFFICULTY	⚙
DURATION	⏱ ⏱ ⏱
REUSABILITY	◍ ◍ ◍
TOOLS & MATERIALS	Paper, pens, action figures, six-sided dice, LEGO bricks

Buildrz (my name for the generic game) is an open-source, build-it-yourself board game for GeekDads to build and play with their kids. The point is, rather than running to the store and buying a game based on someone else's ideas, you can take the idea of a board game and add your own themes and imagination to make it your own.

The idea of a board game is very simple: It is a journey from one place to another, based upon some randomness (dice roll or spinner), with challenges (tasks to overcome, strategies applied against or by your opponents), all dressed up in a motif or idiom to evoke the imagination. The Buildrz game deconstructs the board game to those bare bones and lets you create your own theme and rules.

NOTE: Full instructions with printable boards and cards in various file formats are available on the Web site for this book, www.geekdadbook.com. There are also forums where players can suggest their own modifications to the game.

What's most important about the Buildrz game is what you do with it. At its basic level, it's a fun little game you can throw together with parts you already have, and spend a few enjoyable hours playing. But it can also be a project for the whole family to build together and come up with new themes, new cards, new tweaks to the game play that make it all your own. Maybe it will even become a family tradition of yours: You'll take it along on trips to the family cabin or bring it out for entertainment when the power goes out. Then it's no longer my game, it's yours.

BUILDING THE GAME

First you need to make the game board and playing cards. To make a small board, you could draw the board out on an 11-by-17-inch piece of paper, though that might be tight. It would be better to tape a number of sheets of paper together, or use butcher paper, to make a larger board. Alternately, you could print the board out from the file available at www.geekdadbook.com to a very large size, broken out onto multiple sheets of paper, and tape them together.

HELPFUL IDEAS: A pool table, Ping-Pong table, card table, or other large working space is a good place to set up a large-scale version of the game. One excellent resource for large sheets of paper is construction projects. Old construction drawings have the plans printed on one side, and blank white spaces on the other, which make for great drawing. Using a yardstick or other long straight-edge is handy for segmenting the board.

Then you draw the following on the board:

- A Home circle in the center of the board
- Around that is the Inner River. This is a metaphorical river, and depending upon the theme you choose for your game, it

could be a force field, a gorge to be spanned, or the mystic space between worlds to be bridged.

- A ring around the Inner River
- On the ring is the Inner Path, with twenty-four spaces, four brown bridge Abutments to the Home circle, and four brown bridge Abutments from the Outer Path. Four of the spaces on the Inner Path are Toll Spaces (shown in yellow), and four are Card Spaces (red or green), which are explained in the rules.
- Around this is the Outer River.
- Around the Outer River is the Outer Path, comprised of thirty-six spaces, including a yellow Toll Space adjacent to each of the outer bridge Abutments, and eight (red or green) Card Spaces.
- From the midpoint of each quadrant of the Outer Path are the Trails that lead to each player's starting space. There are thirty spaces along each player's Trail to the Outer Path around the Outer River, including ten (red or green) Card Spaces. This is a guideline number, based upon the idea that the roughly average roll of a six-sided die will be a three (actually 3.5), making for about ten turns to get from the Start to the Outer Path. If you want to use a different die, or make a quicker or longer game, you can play with these numbers as you like. Also note, if you have more or less than four players, you could build a board with more or less than four Trails. Just keep in mind the symmetry that helps create the balanced game play.
- Every third space along each Trail is a Card Space, where, if you land, you pick up a card from one of the two decks. The spaces alternate green for Defense Cards and red for Offense Cards.

- There are also numbered Number Spaces along each player's Trail that will play a part in special moves during the game. Counting from the first space on the Trail just off the Outer Path, and working back toward the Start of each Trail, put "1" next to the first space, "2" next to the fourth space, "3" next to the ninth space, "4" next to the sixteenth space, "5" next to the twenty-fifth space (notice a pattern?), and "6" next to Start. Do this for each Trail. Make a Deck Space on either side of the board for the Offense and Defense cards.

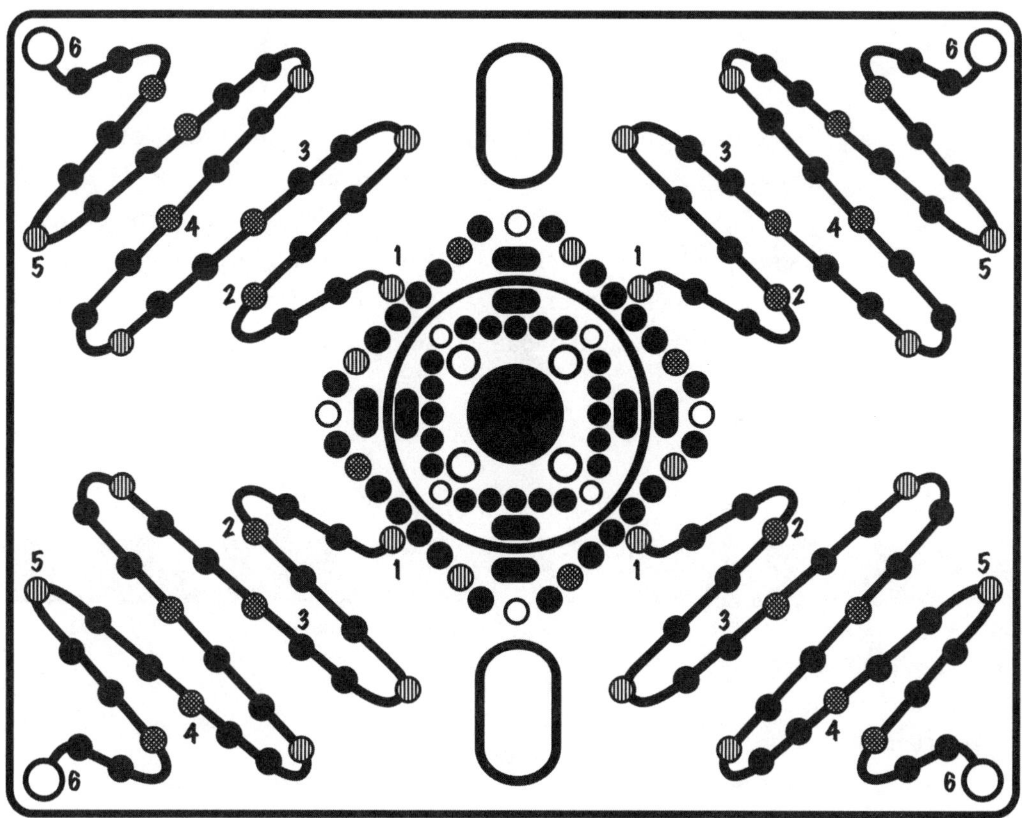

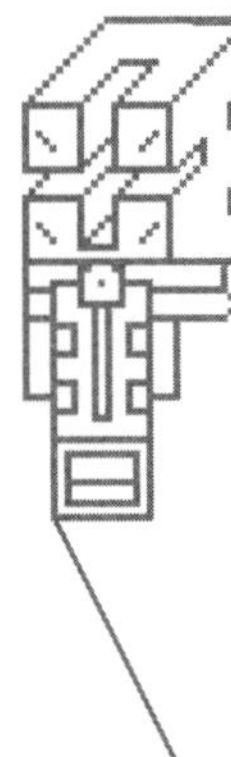

Helpful Hints

When you draw a large-scale board, it's a good idea to do everything in pencil first, using rulers and water-bottle caps for most of the lines and shapes. Then go back over everything with a black Sharpie/Biro, and finally color the spaces in per the directions. If you print it out, it's just a matter of lining things up and taping the pages together. Another cool idea is to use cloth pens and draw the board on a white bedsheet that can be folded up and reused.

Now you need the game cards—red for Offense, green for Defense. These can be printed on paper and cut out, or written on 3-by-5-inch index cards, and a colored dot placed on the backs to identify which kind they are. The following are suggested cards and quantities, but once again, you can tweak the game to your own style and idiom of game play by adjusting the cards.

Offense Cards (Red)

TEXT ON CARD	NUMBER OF CARDS
Move Ahead 1 Space	12
Move Ahead 2 Spaces	6
Move Ahead 3 Spaces	2
Take an Extra Turn	3
Move to the Next # Space	2
Pick/Add a Build Piece	4
Roll and Move to that # Space	1

Defense Cards (Green)

TEXT ON CARD	NUMBER OF CARDS
Move Back 1 Space	12
Move Back 2 Spaces	6
Move Back 3 Spaces	2
Lose a Turn	3
Move to the Previous # Space	2
Lose/Remove a Build Piece	4
Roll and Move to that # Space	1

The cards are the place where the game can be most significantly customized. If you come up with a theme for the game (say, space travel), then the cards can help portray that. "Move Ahead" becomes "Warp Ahead." "Take an Extra Turn" changes into "Time Jump to Your Next Turn." That sort of thing. And new rules and strategies can be added. How about a card that transfers ownership of a bridge or causes it to be removed?

PLAYING THE GAME

Now that you've built the thing, you're just about ready to play. Mix the cards up and put them on their respective places. Each player should pick a playing piece. Have fun with these! Use action figures or, especially good, LEGO minifigs. Make sure your LEGO bricks are in a bowl for picking, and don't place it too close to the snacks, just to be safe.

The Rules

THE GOAL:

Be the first Buildr to cross both rivers and make your way to the Home space.

PLAY PIECES REQUIRED:

- A token for each player—may be a toy, minifig, action figure
- One six-sided die
- A bowl of assorted LEGO pieces (using all smaller pieces will make for a longer, more challenging game) that will be used to build the Bridges to carry you Home

GENERAL CONCEPTS:

- High roll on the die for first player. Play continues to the left.
- Each player starts in a corner and travels down his own Path toward the Outer Trail, based on die rolls.
- Once at the Outer Trail, pieces move clockwise around the ring.

DESCRIPTION OF A TURN:

1. Player declares his Buildr action—the player may take a LEGO piece from the bowl, add a piece to a current build, set a complete bridge if he is on a Toll Space, or add a piece to repair an already set bridge. He may also choose to pass on his Buildr action.

2. Player may play a Buildr card from his hand, if desired. Only one card may be played per turn. If a card is played, either for the player or against another player, its instructions are now followed.

3. Player rolls the die, then moves the number of spaces indicated, either up his Path toward the Outer Trail, or clockwise around the Outer or Inner Trail. Player takes a Buildr card if he lands on a Card Space, and adds it to his hand.

4. Player may play a Buildr card from his hand, if he did not do so in step 2. If a card is played, its instructions are now followed.

4\. Turn ends, and play moves to the player on the left.

BUILDR CARDS:

- Buildr cards may be played either before the die roll or after the player moves in a turn.
- Buildr cards may be played for the player himself or played on another player.
- The instructions on a Buildr card must be immediately followed by the player who draws it.
- Buildr cards that indicate "Move Forward" or "Move Back" will work at any time.
- The "Take an Extra Turn" card causes the card recipient to take another turn immediately following the end of his current active turn (if one is active) or at the end of his next turn (if he is not in the middle of a turn when the card is played).
- The "Move to the Next # Space" card works only on players who are still on their Trails.
- The "Move to the Previous # Space" card works on any player at any time. Players on the Outer or Inner Path will be sent back to the #1 space at the end of their Trails.
- The "Pick/Add a Build Piece" card effectively grants an additional, immediate Buildr phase for the recipient. He may pick another piece from the pot, add a piece to a current bridge build, or add a piece to an already set bridge. He may not set a bridge.
- The "Lose/Remove a Build Piece" card may have one of the following three effects, determined by the player using it: He must put a loose Buildr piece from his collection back in the pot; he must remove a piece from a current bridge build; or he must remove a piece from a currently set bridge,

thereby damaging it and making it uncrossable until repaired.

- The Roll Cards will work at any time, on any recipient (including the player who plays it). The recipient must immediately roll the die and move to that number space on his Trail.

SPECIAL SPACES:

- Card Spaces: These spaces, either green or red, indicate that a player who lands there must draw a card of the appropriate color.
- Numbered spaces: When a Roll Card is played, the recipient must roll the die and immediately move to the space with the corresponding number on his path.
- Toll Spaces: Landing on the yellow Toll Spaces allows bridges to be set. They can be reached only by an exact die roll or through use of a card. If a player has a bridge ready when he lands on a Toll Space, he may set the bridge on the adjacent brown Abutment in preparation to cross.
- Bridges: To cross a bridge, the player must have landed on the Toll Space on the outer side the previous turn, or by card prior to the die roll on a current turn. Each bridge represents three spaces: one for the outer Abutment, one for the bridge span, and one for the inner Abutment.
- Once a player has crossed the outer bridge and landed clearly on an Inner Path space, he cannot be forced back over the bridge by means of a "Move Back" card. He can be sent back to an outer space only via a Roll Card or the "Move to the Previous # Space Card."

STRATEGIES:

- Use of the cards, either offensively or defensively, is very important, especially for reaching Toll Spaces or keeping other players from doing so.
- Players may set their own bridges or wait for others to do so and then use theirs. Any player can use any bridge.
- Bridges may be "damaged" through use of the "Lose a Build Piece" card, making them uncrossable. Bridges can subsequently be repaired only by the owning player, by adding one piece per turn at the start of a given turn.

Other Game Play Issues/Ideas

In the end, the game will be what you make it. You'll need to settle on what's fair for the size of bridges versus the size of the toys you use for player pieces. In general, I suggest the playing piece must be able to stand by itself on the three spaces each bridge represents. And the bridges must span the river, with a foundation resting on each Abutment space. Of course, when people are competing in a game, they will press the rules as hard as possible for an advantage, but fair play should always be encouraged to win out.

Possible ideas for themes/genres to customize the game:

- *Star Wars:* Come up with a microplot that fits into the *Star Wars* universe. Maybe the players are four groups searching for a lost world that holds ancient Jedi secrets that could help overthrow the Empire. The rivers are galactic barriers that must be overcome, and the bridge pieces are actually special technology. Customized cards could mention jumps to light speed, information gleaned from Bothan spies, and upgrades to your ships made by your trusty droid.

- *Lord of the Rings:* The game could be a quest from the deep and rich history of Middle Earth, where the four races—Hobbits, Humans, Elves, and Dwarves—are all trying to discover a way to the lost island of Númenor. Cards could represent magic bonuses or detrimental spells cast by Morgoth or attacks by the Orcs.
- A Spy Game: Maybe each player is a superspy—James Bond, Jason Bourne, and so on—all trying to recover a vital piece of intelligence hidden in a secret place.
- Or *Battlestar Galactica, Chronicles of Narnia, Star Trek, Harry Potter* . . . whatever you and your family love!

An Even Cooler Idea!

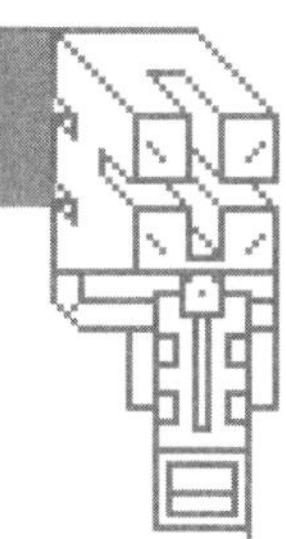

I've given you the basics to create a playable game. But you can do more! Adjust the game however you like. Print the board on 11-by-17-inch paper or cover your pool table with butcher paper and draw it out BIG. Even consider using cloth pens and drawing the board on a white bedsheet that can be folded up and reused. Add spaces with special properties. Add cards that do more/different things. Use different dice.

Adjust your ratio of larger and smaller pieces to make the game more playable for younger children or more challenging to older ones. A reasonable bridge can be made with three pieces, if they're of the right size. But for a longer, more inventive game, the use of no LEGO bricks over two studs by eight studs would force some interesting construction and would result in more time spent on the Outer Path and individual Trails.

If you have LEGO bricks and LEGO minifigs (the little people in LEGO sets), use those. But K'NEX or Lincoln Logs or other generic building sets are also cool for the building part. And then have fun using other toys you have lying around. The first time we play-tested the game, there was a *Doctor Who* figure and the Flash in the mix.

Even better, use whatever theme for the game you want. It could be fantasy (create your own *Lord of the Rings* epic quest) or science fiction (a race across space to claim a new home world for your species) or the ancient world (be the first to bridge the great rivers and claim the throne). Whatever you want!

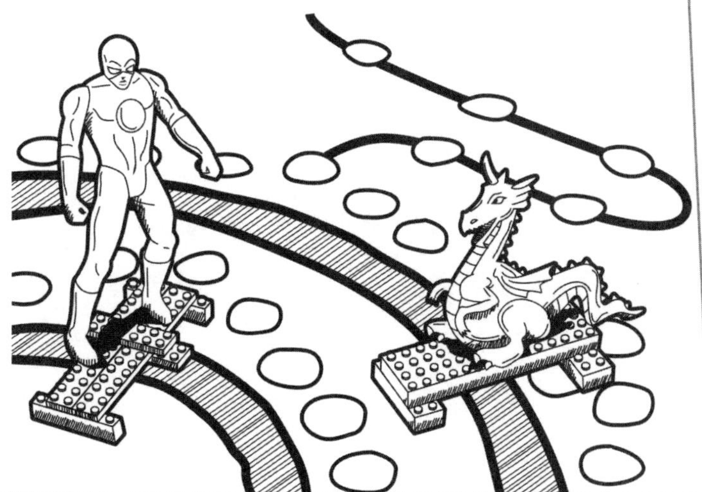

Electronic Origami

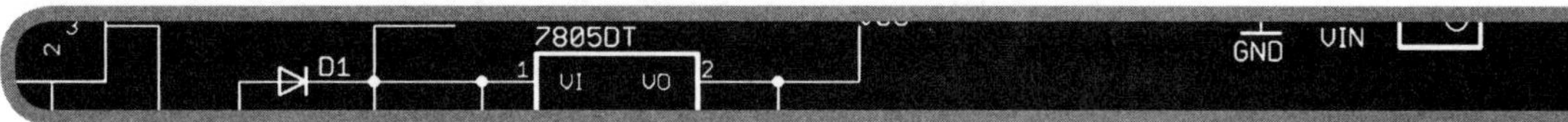

Origami is an artistic tradition dating back at least 1,300 years (and probably more), and while it's steeped in the naturalistic aesthetic tradition of Japanese culture, it has held an appeal for geeks as well. Perhaps it's because of the link to Japanese culture. After all, geeks have a passion for manga and mecha and all things ninja. Maybe geeks appreciate the balance of the technical and the artistic. Case in point, I was the "president" of the Origami Club in my high school, and all the members were my buddies from playing D&D and taking AP Physics.

So origami can be something really fun to share with our kids, especially when they are younger. It's about the least expensive art/craft you can try, and it involves enormous creativity and imagination. And if your kids balk at the idea that folding paper into animals can be cool, just tell them to think of it as making their own action figures, and promise you'll act out Pokémon battles with them when they're done.

But how can we make origami even geekier?

I was browsing the aisles at my local electronics warehouse one day, looking at parts and pieces, and I noticed a very interesting item called a CircuitWriter pen. If you remember those glitter pens that everyone loved to use in junior high school, this is the same idea. But the material is actually silver, in a suspension of acetone, resin, and a few other chemicals with big names. You can use the pen to draw basic electrical circuits or fix broken traces without having to etch or solder; its ink works just like the thin conductive material on a circuit board, and will conduct electricity.

That got me to thinking: What else could you draw on to make a circuit? What about paper? Could you draw a circuit on paper and, say, run an LED from a battery? And, if you could do that, what could you then do with the paper? All of which led me to this project.

PROJECT	ELECTRONIC ORIGAMI
CONCEPT	Trace circuits onto paper, then fold them into interesting shapes and light them up.
COST	$–$$
DIFFICULTY	⚙⚙
DURATION	☼☼–☼☼☼
REUSABILITY	⊕⊕
TOOLS & MATERIALS	CircuitWriter pen (or equivalent), paper, CR2032 battery, LED, tape, large paper clip

This project will introduce you to the electronic origami concept—we're going to keep it simple and build a box with an LED. If you're creative with this idea, you could come up with faux tea lights for decoration, or even emergency lamps.

You can use a regular piece of 8.5-by-11-inch letter paper trimmed down to a square: Fold one corner over diagonally to the opposite edge, and then cut or use a straight-edge to remove the excess section of paper. This results in a fairly large box (about 4-inch square), so once you master the fold and the circuit drawing, you may want to scale down to smaller sizes, which will actually help the circuit—shorter electrical paths means less loss of power to resistance. You may also want to play with different types of paper to see which hold the current lines better. More absorbent papers may require thicker lines.

MAKING YOUR LIGHT-UP BOX

STEP 1: Build your box, based on the instructions in the illustration. Use a straightedge to get good creases on your folds.

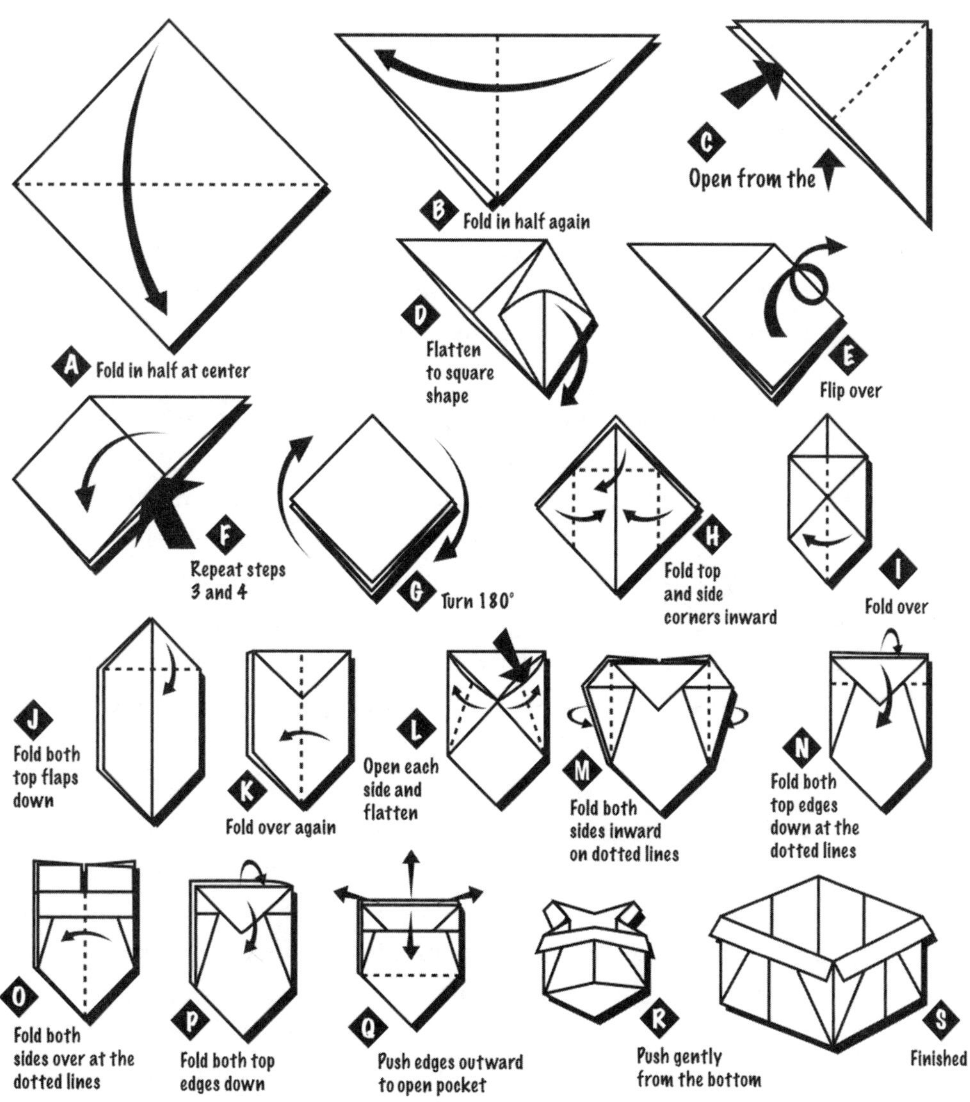

STEP 2: We have to identify where the path of the circuit lines are going to go, and this will take a little careful tracing. Take a pencil and draw a small dot at the center of the inside bottom of your box to identify where the LED is going to sit. Now choose one of the corner sides of the box where the battery will slip in. You'll notice there's a pocket of paper on either side. Insert the tip of your pencil about halfway from the top of the corner and rub it around a bit to make marks on both sides of the paper in the folded pocket.

STEP 3: Now come the electronics! Carefully unfold just the corner side of the box where the marks are by lifting the adjacent top flaps and expanding the folded-over pockets at the corner. Where you slipped your pencil in, you should now see two distinct marks on either side of a fold—when folded, they face each other. These will be the contact points for either side of the battery.

STEP 4: Use the CircuitWriter pen to trace out the two circuit lines. At each of the contact points on the corner, draw a pea-size circle of the conductive ink, and then draw a line down to the floor and in toward the center. At a spot just to the side of the center, make a good pea-size circle of circuit material to end your line. These will be your positive and negative "wires." They should not cross each other, but otherwise the path from the corner battery contact to the floor, where the LED will connect, is up to you. Just keep it relatively short.

STEP 5: Let the page dry (use a hair dryer or fan for quicker drying). Check the lines for continuity and fix any thin spots. Once the page is completely dry, you can do a test run: Hold the LED with its leads touching the contact points in the middle of the page, and then take your battery and carefully fold it into the crease between the other two contacts. Make sure you have your positive battery side feeding to your positive LED lead. If all is right, you should see light.

STEP 6: To finish it off, fold your corner back together.

STEP 7: Looking inside the box, you can see the circuits you traced. Take your LED and, using a little tape, affix it to the center of the box with one lead on each contact. Where the circuit lines vanish into the corner folds, slip your CR2032 battery into the fold, positive side to the positive contact, and negative to negative. To get the battery to fit well in the pocket, you may need to use an X-Acto knife to slit the paper along the adjacent fold so you can then slip the battery in from the outside, under the top flap. Then you can hold it in place with a paper clip.

The LED will light up, and you have built your first piece of electronic origami!

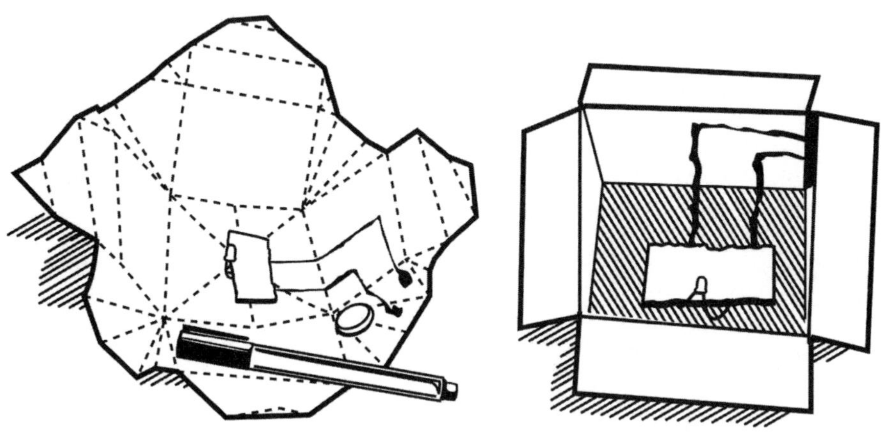

This is just the start, though. The variety of origami patterns available on the Internet is nearly limitless. You can fold a dragon and make a flaming mouth with a red light! You can even make paper airplanes with working landing lights! Anything is possible when geeky parents and their geeky kids work together.

Cyborg Jack-o'-Lanterns

and Other Holiday Decorations for Every Geeky Household

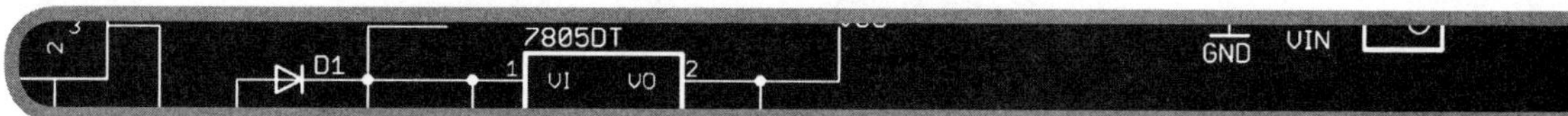

Halloween is a big day for geeky parents and kids. Between coming up with cool homemade costumes and spooky decorations, it's one holiday when you can really get creative and have some terrific fun. The most traditional project of Halloween, of course, is the jack-o'-lantern (JoL). But for a truly geeky family, after a while a simple cutout scary face with a tea candle inside just isn't exciting enough anymore. This chapter will help you turn a tradition into a high-tech project with real geeky appeal!

The winter solstice holidays are also a particularly family-focused time for most households, and geeky households are no different. Any GeekDad worth his salt will do his best to re-create the best parts of National Lampoon's Christmas Vacation where it comes to decorating, and many of us do our best to incorporate technology into the outdoor decorations.

But what about the indoor decorations? Whether it be Christmas, Hanukkah, Kwanzaa, or Festivus that we celebrate, we have to assert our geeky identities! These ideas are great for geeking up the holidays.

PROJECT	GEEKY HOLIDAY DECORATIONS
CONCEPT	Ideas to geekify your holiday decorating
COST	$-$$$
DIFFICULTY	⚙⚙
DURATION	☼-☼☼☼
REUSABILITY	◍◍◍
TOOLS & MATERIALS	A pumpkin, lights (LEDs or other), Arduino board for programmability, motor parts, robot parts, an MP3 player, battery-powered speakers, LEGO bricks, the LEDs and batteries from the Fireflies for Every Season project on page 117, assorted cords and other formerly useful technical accessories

HALLOWEEN

This project isn't so much a single project as a series of ideas for creating JoLs that will make you the talk of your neighborhood. There are a lot of special carving kits available these days to get really creative images on your JoLs—not just faces but all kinds of structures and words. Many of them involve carving away the outer skin and some of the fruit so that lighting inside backlights the imagery. Those are pretty, but they're not what we're after here.

Rather, we're going to go back to the core idea of the JoL—the carved face—and build from that with three dimensions that add life to any JoL: light, sound, and movement.

Carving Your Pumpkin

Many special JoL carving sets come out around Halloween, and many special carving tools get sold. They are, as famous food geek Alton Brown might say, pathetic unitaskers that will get used once and then left in a drawer for a year. Save your money and use the tools you already have. And when I say tools, I mean tools.

While a big serrated knife is okay for the heavy cutting, it's really much faster and more satisfying to use your handheld saber saw! And if you want to do interesting cuts that don't go all the way through the meat, how about a router? You can set it to a depth that won't go all the way through, and then carve out shapes and spaces where you can add some of the items suggested below in the other sections.

And though the traditional JoL face is all well and good, power tools can make other ideas much easier to achieve. For instance, instead of cutting the mouth like a traditional smiley, try a power drill with a ¼-inch bit, and drill a series of holes to look like the grille on an industrial speaker.

Just make sure that, as you would with kitchen tools, you clean up your power tools properly. You don't want to come back to your drill in a couple weeks to find moldy pumpkin, do you?

Light

The classic construction of a JoL includes a candle in the middle of the pumpkin. You can also now use a battery-powered light instead, but most people don't think bigger than a single light. For our JoL, let's use the programmed BlinkM unit we first tried in the Cool LEGO Lighting from Repurposed Parts project (page 191). Please refer to that project for how to program your BlinkM unit, but this time make the pattern something eerie, like a slowly shifting medley of darker colors with the occasional bright flash. Once programmed, it can go inside your JoL just like a candle would, and it can be powered from a DC converter or a battery pack. Other unique but easy sources you could use are flashlights with colored plastic wrap over them, or any of the strobes available around the holiday.

But you can do more with lights. Perhaps the BlinkM is intended just to give a glow from the mouth of your cyborg JoL. For the eyes, get a pair of LED key chain flashlights and cut the eyeholes so they fit just flush inside the surface of the pumpkin. Or pick up two (or more) of the tinyCylon LED kits from www.makershed.com. You could mount them vertically or horizontally into cutout eye sockets and time their blinking patterns for a variety of effects, from creepy robot "scanners" to nutty cross-eyes.

Sound

Sound is a great—and relatively easy—dimension to add to your JoL. The cheapest route is to find an old MP3 player—come on, you know you have a two-generation-old iPod or even one of those thumb-driven el cheapo kinds lying around somewhere. Add that to a battery-powered speaker or (if you're already using external power for your lights) a computer speaker that will fit inside the pumpkin, and you're nearly set to go. Dredge up a good spooky soundtrack and some eerie sound effects, or make yourself a playlist of Halloween songs, and you've got your audio.

To take it one step further, you can create a live version. Put a wireless speaker inside the pumpkin and hook it up to your laptop. Use a mic and some kind of voice-modulating software to produce a really creepy voice, then watch as people come to the door and comment to them about their costumes. To make it really believable, mount a wireless webcam in the pumpkin as well so you don't have to be peeking out a window to see your victims. The effect is really, really cool.

Sound-activated circuits (you know, what made The Clapper work) are also fairly inexpensive and are available online in a variety of places. Use that to activate the light you have inside the pumpkin (May require programming! Good thing you already learned how in the Lamp project on page 191 . . .). Then the interior light could blink on and off as you speak through the wireless speaker, making the effect of a talking cyborg JoL just that much more awesome.

Motion

To fully realize the cyborg effect, you'll need some kind of movement. Perhaps the lid of the JoL could turn and jostle, or laser pointers mounted on the side of the JoL could rotate when people approach. The possibilities are endless.

Two possible methods for making these ideas work are:

The first method is to rip out the guts of an old RC car (maybe a former LEGO Art Car Demolition Derby vehicle) and use the steering and driving mechanisms.

Or second, and a little more expensive (unless, as a good geeky parent, you already have a set), is to use your LEGO Mindstorms NXT set. You could cut the bottom out of your JoL and have it sit over a wheeled robot base built from the Mindtorms set. Then you can control it via remote or even program the robot to move when it "hears" noises or "sees" movement. How creepy would that be?!

Putting It All Together

You're obviously going to need a fairly large pumpkin if you're going to fit all these features inside. Power tools are a great way to do a lot of the cutting work easily. In my time I've used power drills and saber saws to build my JoLs.

If you're going to have electronics inside, the pumpkin will need to be well cleaned out, and pretty dry as well. Consider putting paper towels on the bottom to absorb latent moisture. And the bottom will need to be as flat as you can get it. To run external power into the pumpkin, simply drill a good hole in the back side to feed cords in. If you need to mount things on the inside sides, try building a scaffolding of taped-together bamboo skewers that pierce the fruit.

In the end, a family's JoL is a very personal creation, and no two are ever exactly alike. If you use any of the ideas in this chapter, please take pictures or some video and post them in the forums for this project at www.geekdadbook.com. I'd love to see them and share them with the community.

WINTER HOLIDAYS

As we've already discussed, LEGO is a geek's best friend, especially when it comes to building things yourself. So, if you're going to make your own decorations, why not break out the bricks?

You can save the money you might be tempted to spend at a certain greeting card store on tree ornaments that look like the starships from your (our) favorite science fiction television shows. Build something cool instead!?

LEGO Holiday Tree

A really easy and very repeatable project is to build a small LEGO holiday tree and then use the battery and LED concept from the Fireflies project on page 117 to add tree lights. Indeed, because LEDs don't get hot, you can get a very cool effect by their shining from within the colored, transparent LEGO bricks:

1. Start with a 2-by-4-inch brown brick as the base, and then stack the following brown bricks onto it in layers, centered on the middle quad of studs: 2x2, 2x2, 2x4, 2x2, 2x2, 2x2. You now have the trunk of your tree.

2. At the second 2-by-4 brown brick (or the fourth level), attach a 2-by-4 green brick under either side where the brown brick overhangs the smaller bricks below it (one green 2-by-4 brick on each side).

3. Attach a 2-by-2 green brick on top of each of the 2-by-4 greens you just attached, so that the 2-by-2s are now adjacent to the 2-by-4 brown on either side.

4. Attach a 2-by-3 green brick on either side so that it sits on the two exposed brown studs and the four exposed green studs of the bricks you attached in step 3.

5. Stack two 2-by-2 green bricks on each side atop 2-by-3s from step 4 so they flank the two stacked 2-by-2 brown bricks that form the top of the trunk. The top of your green bricks should now be flush with the top of your brown bricks, leaving a 2-by-6 surface.

6. Stack two 2-by-4 green bricks on top, centered.

7. Add one 2-by-2 green brick on top of those, centered.

8. If you have it available, use a single 1-by-2 green brick with a hole in the middle (common to Technics sets) at the very top, as the mount for your LED.

9. At the fifth, seventh, and ninth brick levels, add a 1-by-2 transparent colored brick on the outside on either side. This should complete the treelike appearance and will look like glass ornaments. If you have them, put colored (perhaps, say, red and/or green?) studs on top of the tenth level on either side as well.

10. Slip the LED of your firefly into the hole on the top brick, tape it down from the back side, and light it up!

If it wouldn't violate your level of orthodoxy, you could apply this idea to a menorah instead (or as well). Build a LEGO menorah, with each of the cups designed as a repository for one of the fireflies. This applies to the traditional lighting for Kwanzaa, too.

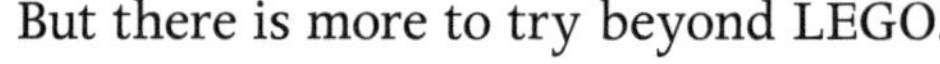
But there is more to try beyond LEGO.

Geeky Wreath

One very common decoration, whether it be Christmas or its cultural predecessor, the winter solstice, is the wreath. From the Wikipedia entry under "Advent Wreath":

> The ring or wheel of evergreens decorated with candles was a symbol in northern Europe long before the arrival of Christianity. The circle symbolized the eternal cycle of the seasons while the evergreens and lighted candles signified the persistence of life in the midst of winter.

While these were horizontal wreaths (which eventually became the Advent wreath), at some point they also went vertical and were hung on walls or doors as decoration. These days, there's quite a lively business around the design of decorative wreaths.

For our wreath, however, rather than evergreen branches and flowers, we're going to be a little more environmentally sensitive and repurpose some materials all of us good techie geeks have lying around.

Somewhere in your house, there is a box or basket or plastic tub, and in that tub are cords. They may be extra USB cords, or Fire Wire, or serial, or even parallel. They may be AC adapters or power cords from old computers or monitors or the proprietary data cord that the 2-megapixel camera you bought eight years ago used to connect to your TV to display pictures. There is probably even a handsome selection of component, composite, DVI, and other AV cords, all piled together for that day when you KNOW you'll need them. Which is why you've kept them all.

The first step toward healing is admitting you have a problem.

The next step is to take them all out and braid them into a holiday wreath.

Depending upon how adroit you are at actually braiding odd-size cords, you may wish to cheat a little by going to your local crafts store and purchasing a wire or wood frame that you can braid the cords around. You can use zip ties or even just plain tape to keep them on (an artistic application of electrical tape could provide just the right look).

The important thing is not to make it look organized and evenly assembled; a bit of chaos is good in this case. And if you really want to go that extra step, don't stop at cords. Just as pretty front door wreaths can have ornaments or other decorations on them, you can do the same with the random technical detritus accumulating around your house: old hard drives, computer mice, webcams, routers, fans, speakers, whatever. Hang them from the wreath and spread the geeky spirit. Add one string of real holiday lights to bring it to life, then hang it on your door and impress the neighbors.

Windup Toy Finger Painting

Like coloring, finger painting is one of those quintessential childhood creative joys. Well, joyful for the kids, but less so for the parent or teacher who gets to do the cleanup afterward. At the end of the day, however, it's all about letting your children express themselves so you can end up with an abstract piece of art to hang on the fridge or your cubicle wall for a few months.

But as geeky parents, we're always looking for a new twist. This is a project perfect for your younger kids, and just right for a summer day or weekend when you need to kill time in a creative way.

PROJECT	WINDUP TOY FINGER PAINTING
CONCEPT	Make abstract art using the cheap windup or battery-powered toys you have lying around.
COST	\$–\$\$
DIFFICULTY	
DURATION	
REUSABILITY	
TOOLS & MATERIALS	Butcher paper, paints, self-powered toys

On a creative scale, this projects sits somewhere on the line between classic finger painting and Spirograph. The simple idea is to use whatever windup toys you have as the brushes for your painting. But you can take it even further.

The setup is easy. Butcher paper, available at your local craft store, is the easiest canvas. Lay it out on a floor or table to the size you want for your final artwork.

Now, and this is key for this project, you need bumpers. Set up barriers to constrain the toys so they don't go wandering off the canvas while they're working. Carpenter's levels, rulers, and lumber will all work, depending on the required size.

Toy choice is important. You may decide just to use whatever you have lying around. There are quite a lot of windup toys in fast-food kids' meals these days, so you probably have some of those lying around. But if you want to make this more a techie project, consider looking for the various toy mini robots that have simple sensors on them so they can sense and avoid walls. These will give you maximum painting bang for the buck.

Unless you're worried about the work of art lasting for all eternity, pick up the cheapest paints you can find at your art store, in a variety of bright colors. Pour a bit of each color into low cups or saucers, and dip the feet or treads of one of the toys into it. Wind it up, then set it down on the paper to start doing its stuff. Repeat with different toys and different colors. Hilarity will ensue.

Of course, one significant concern here is possibility of destroying the toys—you'll need to wash the paint off the toys if you want to keep them. This is doable with most inexpensive paints; just be careful not to get water in the mechanisms.

An Even Cooler Idea!

If you want to make the artwork a bit more permanent than something to magnet to the fridge, consider using spray glue to laminate it onto a piece of board and then putting a spray poly coating or two over it.

Simple fun for an afternoon.

Create a Superhero ABC Book

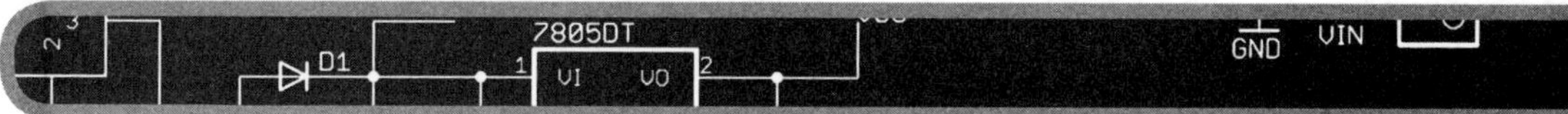

I hate having to hear my kids say, "Are we there yet?" when we're on a car trip. So I always do my best to keep them entertained. (Okay, fine. The DVD player helps a little, too. . . .) One time, I decided to play a word game with them. Since we'd been playing with lots of superhero figures recently, I figured I'd work with that.

"Let's play a word game about superheroes," I said. This got their attention and they stopped fidgeting long enough to show their curiosity. I presented the game: We would go through the whole alphabet and name at least one superhero whose name starts with each letter. *A*. Aquaman. Angel. Apocalypse. Soon we were off and rolling. We thought of Batman, Blue Beetle, and Beast for the letter *B*. Sure, ol' GeekDad had to help jog their memories at times, but they were sucked into the game quickly and we still play it all the time. Sometimes we use superheroes; other times we try to come up with Super Mario characters.

But the game got me thinking that it would be cool to create one of those ABC picture books, starring superheroes. I didn't want to make it just for my kids, though. I wanted to make it with them. And that's exactly what we did.

PROJECT	CREATE A SUPERHERO ABC BOOK
CONCEPT	Make an alphabet/picture book at home using as many superheroes as you can find.
COST	$
DIFFICULTY	⚙
DURATION	☼–☼☼
REUSABILITY	⊕⊕⊕
TOOLS & MATERIALS	Construction paper (about 15 sheets, different colors), hole puncher, yarn (as crazy and colorful as you can find), glue stick, scissors, computer, printer, good search engine

[This project was first developed for GeekDad.com by Andrew Kardon.]

STEP 1: Calling All Heroes

Collecting the images is your first (and longest) step. You'll want to get a number of different pictures for each letter. We used one side of a piece of construction paper for each letter. But you could easily use two pages as a spread, which means you'll need more pics. The easiest way is just to jump online and start searching for superheroes whose names begin with each letter.

There's no harm in admitting you need some help, so a number of online superhero dictionaries came in handy when I was stuck for some letters or additional characters. These were some of the most helpful for me:

The Superhero Dictionary (http://shdictionary.tripod.com)
Comic Vine (www.comicvine.com)
Marvel Universe: The Official Marvel Wiki (http://marvel.com)

Find a good picture of each character (have your kids help pick out which one they like), save them to your desktop, and then print them on a color printer. You may need to do a little work to resize some of the images, but in the end you'll have a big stack of printed pictures of colorful, spandex-wearing superheroes. And as long as you're not planning to sell the book, you don't have to worry about any copyright issues.

After they're all printed out, grab your safest kiddie scissors and have your guys go to town cutting out all the characters. The big rule I had was that just about anything goes. If my kids wanted to keep a background on an image, that was fine. If they wanted to cut out only a character's head, that was cool, too. This was their book, so whatever they wanted was fine with me.

STEP 2: Punch Out!!

Once you've got all your images ready, it's time to get the book ready. So prep the book itself by taking thirteen sheets of construction paper and punching three holes along the left side. Punch the holes before any real work is done, or you may end up punching holes over a picture or letter. Once that's done, have your kids use crayons or markers to start writing one letter of the alphabet at the top of each side of each page. Just be sure the color's dark enough to show up.

My kids also wanted to include a front cover, back cover, title page, and dedication page. So I made sure to include four more sheets in the process.

STEP 3: Break Out the Glue Stick!

Now comes the fun part. Start making piles of your superhero pictures, grouping them by letter. When you have all the *A* name superheroes together, grab a glue stick and start attaching them to the *A* page of the book. Since our pages were double-sided, we did half the alphabet first and let those pages dry before flipping them over and finishing off the second half of the alphabet.

As long as the pictures don't cover any of the holes in the spine, you're fine. We tried to cram as many pictures as we could onto each page, even if some of the pictures had to come off the paper to do so. It just adds a bit of 3-D magic to the fun.

When all the pictures are glued on and dry, have your kids

make a cover. Be sure they include a title and a byline so they can give themselves the proper author credits. A title page is just another excuse to write the title again and fit a few more pictures. Same goes for the dedication page, which gives your little geeklets a chance to act like a real author and dedicate the book to a friend or loved one.

STEP 4: Avengers (and everyone else), Assemble!

The final step is just putting everything together. Literally. Cut out a length of yarn and pull it through one set of holes, and then tie it into a neat bow. Repeat two more times and you're all set. One handy dandy "Superhero ABC Book" ready to read, or to be read to you by your little ones. Just be careful turning the pages, and always have a glue stick handy to re-adhere any pictures that find their way free.

If your kids enjoyed making it, you can very easily substitute any other theme based on their interests. Baseball players, video game characters, cartoons, food, etc. Good luck!

Model Building with Cake

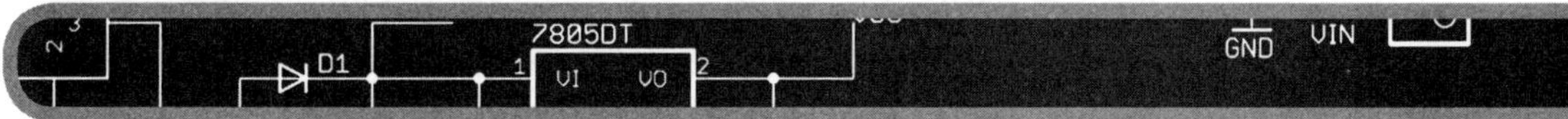

Model building is a geeky practice par exellence. From snap-tight car and plane kits to Warhammer dioramas or Eiffel Towers made from LEGO bricks or toothpicks, the obsessive drive to represent something large on a smaller scale, out of alternate materials, has been with us for a long, long time.

But there's one kind of model building where two different corners of geekery come together. If you and your kids are food lovers, you can bring a little bit of that passion into the world of model building with this project.

PROJECT	MODEL BUILDING WITH CAKE
CONCEPT	Do building projects, for school or fun, out of cake.
COST	$
DIFFICULTY	
DURATION	
REUSABILITY	
TOOLS & MATERIALS	Cake mix, generic puffed rice cereal, marshmallows, fondant, frosting (mix or canned), food coloring, various cooking pans and utensils

The go-to reference for this kind of project is the hit show *Ace of Cakes.* Star cake artist Duff Goldman and his team build what are, in effect, elaborate models out of cake and cake-related products, and if you notice they all are, to one degree or another, geeks. (Okay, we'll admit, one or two of them come across as hipsters, but we'll cut them some slack for the amazing work they do.)

Building a model of something is a task that falls into the lap of every kid at some stage of their educational career. In California, fourth graders have to build a model of a California mission as part of their graduation requirements. In some school districts in Pennsylvania, students may choose to build a historical model to meet specific graduation requirements. And countless other schools urge their students to re-create historical buildings, places, or events in scale replica form as a means of imparting a deeper understanding of a specific subject or era.

Usually these projects are built with Styrofoam or clay with plastic animals or people on a foam core board and are . . . uninspired. But if we add some creativity, we can achieve something much more interesting and fun (not to mention tasty). We can make our models out of cake.

This project is based on the actual model of a California mission my son Eli and I built for his fourth-grade history requirement. Eli has been a foodie since a very young age—he loves to watch Alton Brown and *Iron Chef America* just as much as any cartoon or baseball game. It was his idea to tackle this model in cake. When my wife and I heard the idea, we beamed with pride.

When you're planning a model out of cake, especially a model of a building, there are three key materials to use: cake, generic puffed rice cereal treats, and fondant. You'll use the cake for most major structures (in our example, the large solid structure of the mission church) and use the generic puffed rice cereal treats for most of the other structures—as it's very easy to mold and cut into custom shapes. Fondant is the covering that makes everything look smooth and finished.

When you're planning your model, you need to decide which materials are going to be right for the different parts of the structure. Basic geometric shapes can be done well with cake, since you can stack a number of small sheet cakes on top of each other and then use a serrated knife to cut and shape it. You should create more delicate or elaborate structural elements like columns or arches out of the generic puffed rice cereal treats since it can be molded almost like a very grainy clay and has some useful structural properties; it can be molded into shapes, like columns and arches, or other decorative three-dimensional features.

A California mission is a fairly simple setup. There is a main church building—geometrically a rectangular prism with a triangular prism on top—and then the friars' quarters and other facilities built off to the side of the church in a square surrounding a courtyard with a garden and fountain in the middle. These side buildings usually have verandas, or shaded walkways with arched openings. Our plan was to build the church itself in cake, and the buildings around the courtyard in generic puffed rice cereal treats, then cover everything in a layer of white fondant, which would end up looking very similar to the white-washed stucco common to the missions.

1. Start by making four cakes, all cooked in simple rectangular baking pans. This can be from the cheapest boxed yellow cake mix you can find, or something nicer, depending on one major decision: whether or not you want to eat the cake. If the final product of the project is going to be a model for a school assignment, it's likely it will sit out for days after completion, meaning it won't be fit for eating and you don't have to worry as much about taste. But if you want to have your cake and eat it, too, it never hurts to upgrade your ingredients.

2. If you are mixing your own icing, do it now—make two

batches of white icing and set them aside to act as cement during the construction phase.

3. While the cakes are baking, mix up five batches of generic puffed rice cereal treat mix (not naming any trademarked brand names here, nosireebob!), also molding each one into a baking pan so that when they set, they can easily be cut into blocks.

Recipe for Generic Puffed Rice Cereal Treats

SETUP TIME: 10–15 minutes
TOTAL PRODUCTION TIME: 30 minutes (per the recipe, but you can stagger production to speed things up)

INGREDIENTS:

3 tablespoons butter
1 bag of 40 regular marshmallows, or 4 cups miniature marshmallows
6 cups generic puffed rice cereal

DIRECTIONS (STOVETOP):

- Melt the butter in a large saucepan over low heat. Add the marshmallows and stir until completely melted, then remove from heat.
- Add the generic puffed rice cereal and mix until well combined.
- Spoon into a greased rectangular baking pan, then put aside to cool and set.

DIRECTIONS (MICROWAVE):

- Heat butter and marshmallows in a bowl on HIGH for 3 minutes, stirring at regular intervals to keep smooth.
- Add the generic puffed rice cereal and mix until well combined.
- Spoon into a greased rectangular baking pan, then put aside to cool and set.

Now that you have a whole bunch of treats lying about, we should build something fast so it doesn't all get eaten!

4. Start construction with the main church building. Since you'll want to stack the cakes, you will need to trim the tops off flat. It's easiest to use an electric carving knife, but a large bread knife works well, too.

5. Stack three cakes, putting a layer of frosting in between each one as a form of glue. With all the cakes stacked, you have a good basic building shape.

6. For the arched top of the church, set the last cake on top, with a layer of icing between it and the third layer. With your knife, trim it into a triangular prism with the high point at the center, tapering off to nothing at the long sides.

7. Next comes your first go-about with fondant. It's an interesting material, like a sugar-based clay that will harden stiff after a day or two. When you start, though, you can roll it out like pie dough—and that's just about how to treat it. For this purpose, you want to roll out a batch (make sure you buy it by the bucket; you can find it at your local craft or baking supply store), getting it down to about $^{1}/_{8}$-inch thick and big enough to lay over the church building.

8. You and your kids will need to move it together and lay it on, very carefully. Once it's on, you can poke and prod it with your fingers and/or small tools (a clay modeling tool works, or even a dull butter knife is good) to get it to fit over the shape. Then smooth it out with your hands until you're happy with the fit and coverage, and trim it at the base. When you're done, it should look very much like white-washed adobe or stucco.

9. The rest is detail work. You can make the sides of the garden and steeple out of rice treats by cutting the approximate shapes out of the pans and then hand-molding and covering in fondant.

10. You can make bushes and trees around the building with rice treats as well, and color them green or brown by painting (or airbrushing) them with food coloring.

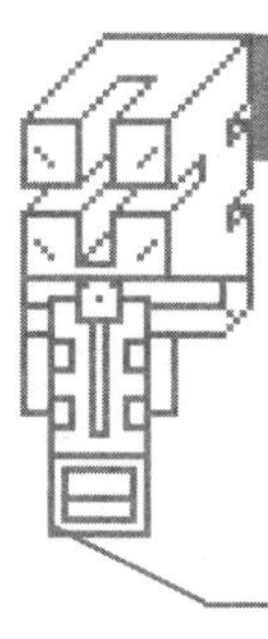

An Even Cooler Idea!

- Make terra cotta roof tiles out of little squares of fondant and paint them with food coloring.
- Populate your mission with historically accurate action figures if you can find them at your hobby store.

This is a very specific example, and if you live outside of California, it may not be quite as compelling a project. However, these materials and methods can be applied to almost any model-building project you can think of, either for school or fun (especially if you want to make an *Ace of Cakes*–style surprise for someone). It's up to you and your kids to use your imagination.

Pirate Cartography

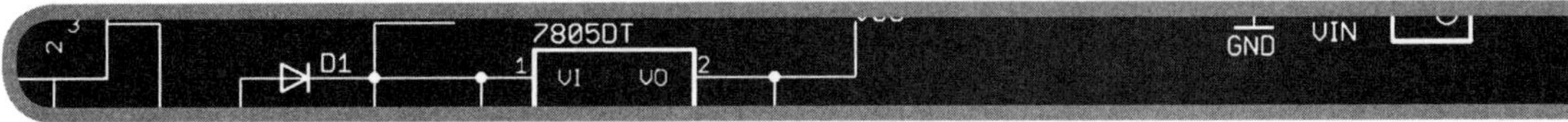

There are very few things cooler to younger kids than pirates (well, maybe ninjas, but that's an age-old conflict we do not wish to delve into here). Even before Johnny Depp's high-profile contribution to the mythology, kids have loved playing pirate since Long John Silver leapt off Stevenson's pages. And as parents, we love helping our kids use their imaginations. We buy our kids eye patches and plastic swords to play pirate dress-up, or we throw them a pirate-themed party. But maybe the coolest way to immerse kids in imaginative pirate play is to set up a treasure hunt. And what does any good treasure hunt need first? Why, a map, of course! And mapmaking most certainly qualifies as a geeky endeavor, whether it be for a pirate party or an adventure campaign.

PROJECT	PIRATE CARTOGRAPHY
CONCEPT	Make an authentic-looking antique map for a treasure hunt or imaginative play.
COST	$
DIFFICULTY	⚙
DURATION	☼–☼☼
REUSABILITY	⊕⊕⊕
TOOLS & MATERIALS	Paper grocery bag, scissors, pencils, markers, crayons, candles

[This project was first developed for GeekDad.com by Russ Neumeier.]

First, you need to create your canvas. If possible, find a paper bag with nothing printed on it. If that's impossible, just use the inside of the bag, which should be blank. To get the largest starting piece out of one back, use scissors or a mat knife to cut away the rectangular bottom of the bag. Then cut up the single seam on the side of the bag. You should be left with a 17-by-38-inch sheet of coarse brown paper.

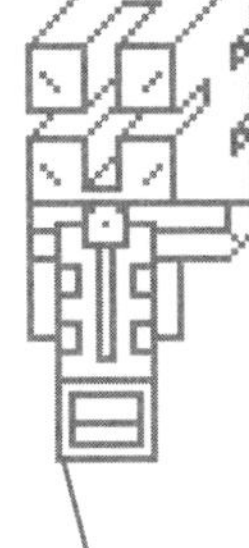

Ideas for Your Map

- Set up a real treasure hunt. Make the map an accurate, though fantastical, representation of your home. Hide things around the house and yard, then give clues to find them, based on outlandish names and drawings you make of the features used on the map.
- A good map is an excellent place to start for storytelling. Make a fantasy map, and then use the imaginary setting each night to collaborate with your kids on an adventure as part of their bedtime story.
- Handmade maps are excellent props for role-playing games!

STEP 1: Sketch your map out lightly, to start, using a gentle pencil stroke. Feel free to try things, erase and try them again, since the bag paper will take erasure well, and every bit of wear and tear will only add to the worn and aged effect of the finished map.

STEP 2: Use markers, crayons, pastels, or even charcoal and watercolors to go over the pencil sketching and give the map some color and interest. Try to decide on standards, such as black for land borders, blue for bodies of water, and so forth. If it's meant to be a pirate map for adventure, less detail and more imagery is good. If you're prepping this for an RPG adventure, you might want to pull out a fountain pen and go to town with the elvish script. Make sure you label key features as appropriate and—most importantly—put an arrow/compass image to indicate North. You might also include a legend and scale, depending on what your use is going to be.

STEP 3: Now that the basic map is finished, it's time to add the style—which means distressing it. In many ways, this is like breaking in a baseball glove. You want to crumple up the map, unfold it, and crumple it again in a new way. You could crumple it up and place it under your sofa cushion before an evening's movie watching. You could drip oil on it, or tea, or beer, or red wine, then dry it with a hair dryer. You should use a candle flame to singe the edges all around, to get rid of the "I was cut out with a pair of scissors" look, and while you're at it, burn a hole or two through it in random places, and then hold it above the candle to get some good soot marks. Candle wax can also add some good staining.

And that's it! You now have a really cool-looking antiqued map for fun and adventure!

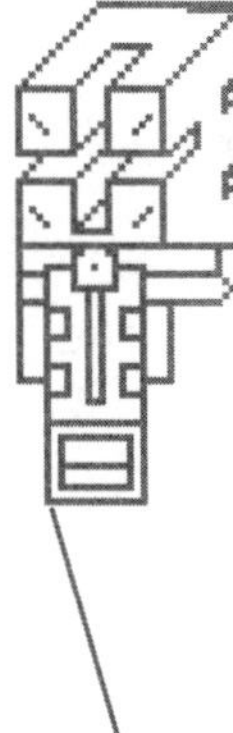

An Even Cooler Idea!

For older kids or RPGs, use the old "real map hidden in a fake map" trick. Make your real map. Then make a second, fake map. But cut out two identical sheets for it. Draw everything on one of the sheets, and then antique both of them up to the point of singeing the edges. Just before doing that, fold up the real map and place it, like a pickle in a sandwich, between the two fake map sheets, then glue the fake sheets together at the edges with white glue. When the glue is dry, singe the edges of the double-thick map. Now let your adventurers loose with the special map, and see how long it takes for them to figure out the secret!

Parenting with Role-Playing Games

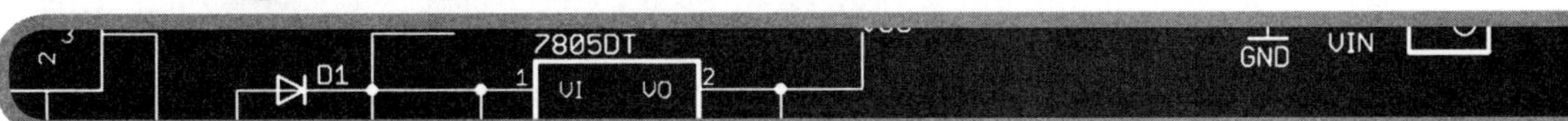

One of the geekiest things you can tell a person is that you play pen-and-paper Role-Playing Games (RPGs). But playing RPGs—especially Dungeons and Dragons (D&D)—was a formative experience for so many of us geeks who came of age through the 1970s and 1980s. While in those days, uptight preachers warned of the devil's influence wrought by RPGs, I don't think you can overstate the value these games have for fostering creative play in kids. I have fond memories of playing RPGs with my friends while we happily snarfed pizza and Jolt and rolled d20s to save our anti-paladins from the breath attack of a white dragon.

Back then, the rules were pretty arcane (quick, what's your THAC0?), and in a way they appealed to all the classic geeky traits, especially depth of knowledge and an obsessive nature. But the game was also intensely social, and many of the bonds made over afternoon expeditions to the Barrier Peaks have lasted a lifetime.

These days, D&D is in its fourth incarnation of rules, and by all accounts, it's easy and fun to play—all the materials are available at your local gaming store (search it out if you don't know where it is; it may even be masquerading as a hobby store; or go online to Wizards of the Coast, www.wizards.com). There is also a whole world out there of similar games based in fantasy worlds, or cyberpunk near-futures, or the *Star Wars* universe, or a present-day Earth with superheroes. The best thing, and the thing that separates them from video games, is that they use the imagination. And they bring people together to play, face-to-face.

Sounds like a perfect thing for GeekDads to do with their kids, doesn't it?

Indeed, depending on your kids' ability to follow rules and their penchant for telling stories, any time in their late single digits, age-wise, is a perfect time to get them into RPGs. First, you get to work with them to build a character—some kind of hero, most likely—that will be their alter ego in the game. And then, you get to lead them through a story—an adventure—either from a book or of your own making. If you're acting as the game master (GM), you even have the ability to modulate the game play so it's just as challenging as it needs to be to keep them interested without ruining their fun. And games like these help teach teamwork, puzzle solving, and even mapmaking.

Beyond just the game play, there are the figures to be painted, the dice to be collected, and even hundreds of licensed novels to be read about the worlds the games are set in. RPGs can be a lifelong hobby that grows with your kids.

But here's an idea: Since RPGs are a way for using mathematics and imagination to turn the life of a created character into a game, why not try applying RPG concepts to managing day-to-day life? Indeed, many RPG players will, at some point in their lives, play characters in an adventure who are based quite specifically on themselves. It's the classic "what would I do if I were magically transported to Middle Earth/*Star Wars*/DC Comics universe?" scenario. Countless players have worked up character sheets for themselves, always slightly overinflating their key statistics (of course I have a high intelligence! I wouldn't be here if I didn't!), and then playing themselves as calm and cool in the most incredible of imagined circumstances. After all, wouldn't every high school junior have the steely nerve to face down a green dragon with nothing but a +1 dagger?

Sure, that's a fun, one-off adventure. But that's still playing a game with friends, or you, as the proud parent, running the adventure for your kids. What if we take it the next step: really turning your kids' lives into an RPG?

Okay, not all life. You're not going to tell your kids to start carrying broadswords to school to deal with the class bully. But the mathematical system of an RPG is meant to create a framework for managing personal growth and achievement for an imaginary character. It creates a balanced reward system for success, and encourages learning and personal planning. What better tools to use in helping your kids handle homework, chores, extracurricular activities, and allowance, all while teaching them to plan ahead and work toward goals? Not many that I can think of, which is why I give you:

PROJECT	PARENTING WITH ROLE-PLAYING GAMES
CONCEPT	Use an RPG-based system for managing rewards and responsibilities as your kids grow up.
COST	$
DIFFICULTY	⚙⚙
DURATION	☼☼–☼☼☼
REUSABILITY	◍◍◍
TOOLS & MATERIALS	Anything you would normally use for playing an RPG. You can do everything on pen and paper, or take it to the computer. You will need dice.

Kids like structure, like having a framework so that they understand what they need to do to receive certain benefits. They also like to have goals to work toward to prove they are growing up and earning new responsibilities. Sounds an awful lot like what you do with a D&D character, huh? So, instead of simple chore charts on the fridge and a weekly allowance, why not turn your kids' duties and benefits into a role-playing game?

THE IDEA

You're the parent, and you are the Game Master/ Dungeon Master (GM/DM). Your child maintains a character sheet with stats, skills, experience, and so forth. Your child earns experience points (eps) for completing regular tasks (e.g., keeping his room clean, walking the pets, washing the dishes) and can get bonus eps for one-shot tasks (an especially good or improved report card, cleaning out the garage, a birthday treat).

At certain point totals, your child can "level up"—gain a promotion representing his achievements and personal growth. At each leveling up, your child earns skill points and attribute points to spend. Skill points allow the kid to learn new skills (how to safely use the lawn mower to mow the lawn, supervised and unsupervised; how to do the laundry) which can then earn him more experience points. Skill points can also give him benefits—like the ability to earn more allowance, watch TV an hour a week, or add an extracurricular activity at school for example—at a measured pace so that your child balances recreational time against taking care of responsibilities. Additional attribute points help raise the bonuses, and are indicative of the kid's personal growth—he is getting more active (adding to physical health, aka constitution), gaining physical agility (improved dexterity), or learning life lessons (developing wisdom).

Complexity can grow as the levels go up (and the child gets older). At the tweens, for example, your child could earn the skill/ ability to have his own cell phone, plus the cost of the plan to pay for it. Your child can also use the character sheet to track money (like how much gold an RPG character has) and budget, so then he can spend it on new items to help him earn more eps or money (the child saves up to buy a lawn mower or power washer to run his own neighborhood business). It's really a matter of how far you and your kids are willing to take the metaphor of the game system. They may, at some point, rebel against the idea of using a game system to manage their lives. Or they may revel in the structure. As with all things relating to the parent/child relationship, your mileage will vary, and you know your kids best.

THE RULES

This system is patterned rather significantly on D&D and a few other RPGs I've played over the years. The key factors for your child's "character" are Attributes, Race, and Class. Using the blank character sheet included in Appendix B (which is also available as a downloadable file at www.geekdadbook.com), help your "player" create his character, thinking a lot about what his own strengths and interests are, and crafting the character in such a way that will optimize the results from doing things he would already be expected to do—yes, help him min/max the character if you like, or keep him balanced if you want to encourage him to do a broader range of activities. Since you'll ultimately control the challenge rolls your children will make and the experience point value of the challenges they face, you can still adjust the "game" based on how they craft their characters at this stage. If they maximize certain attributes in the hope of making successes easier down the line, you can always set a higher challenge rating to balance. Just don't tell them you're doing it—you'd be likely to hear "but that's not fair!"

Character Attributes

ABBREVIATION—NAME

STR—Strength (affects Combat challenges)
INT—Intelligence (affects Magical challenges)
WIS—Wisdom (affects Willpower challenges)
DEX—Dexterity (affects Agility challenges)
CON—Constitution (affects Endurance challenges)
CHA—Charisma (affects Performance challenges)

Challenge Modifiers for All Attributes

SCORE—MODIFIER

6 ....-4
7 ....-3
8–9.......-2
10–11....-1
12–13 ... +0
14–15.... +1
16 .... +2
17 .... +3
18 .... +4

Players start with 12s straight across, and 5 Attribute points to distribute. For every gained attribute point, they can also move an existing point from one Attribute to another. For example, if a starting player wanted to max out his STR, he could use all 5 of his starting points to raise the 12 to a 17, and then move an additional point from INT (making it 11) to STR to end up with an 18. Said player could still move around an additional 4 points, say from INT to DEX, ending up with the following Attributes:

STR: 18 (+4 to all Combat challenges)
INT: 7 (-3 to all Magic challenges)
WIS: 12
DEX: 16 (+2 to all Agility challenges)
CON: 12
CHA: 12

Character Race

Use the classic D&D/LotR races for this to keep it simple.

Human	(no +/-)
Elf	(+10% eps for outdoor/nature challenges, -10% for indoors)
Dwarf	(+10% eps for indoor challenges, -10% for outdoors)
Hobbit/Halfling	(+10% eps for crafting challenges, -10% for athletic challenges)
Half-Orc	(+10% eps for athletic challenges, -10% for crafting challenges)

Feel free to make up your own additional character races based on what your kids may want (girls may enjoy Faeries, boys may want Ogres—or vice versa!), but always make sure the balance is +10%/-10% for the experience from contrasting kinds of challenges they may face).

Character Classes

Depending on the class, the character will earn one additional skill point every third level in a focused area. Skill points can be used to acquire a new skill or to increase the character's rank in an existing skill.

Fighter	+1 skill/rank tagged as "combat" and "solitary"
Warrior	+1 skill/rank tagged as "combat" and "group"
Paladin	+1 skill/rank tagged as "combat" and "spiritual"
Ranger	+1 skill/rank tagged as "combat" and "environmental"
Wizard	+1 skill/rank tagged as "magic" and "computers"
Mage	+1 skill/rank tagged as "magic" and "electronics"
Conjurer	+1 skill/rank tagged as "magic" and "creative"
Illusionist	+1 skill/rank tagged as "magic" and "performance"
Cleric	+1 skill/rank tagged as "support" and "spiritual"
Druid	+1 skill/rank tagged as "support" and "environmental"
Thief	+1 skill/rank tagged as "acquisition" and "resourceful"
Monk	+1 skill/rank tagged as "physical" and "agility"
Minstrel	+1 skill/rank tagged as "performance" and "group"

Again, feel free to add or edit these classes depending on your children's interests. Just stick to the game mechanic of granting one additional skill point every third level in an area tagged with two specific types.

Tags

Tags are ways of organizing ideas. In this game, both Skills and Challenges are tagged with different words that help identify what concepts they apply to, helping sort out what bonuses may figure in determining experience. Ultimately, how Skills and Challenges are tagged is up to you and how you want to run the game. For instance, you may decide that allowing your child to apply his Magic bonus to creating a good science fair project makes sense, but applying his Agility bonus does not.

Skills

In our game, skills represent not only things your child knows how to do, but may also be permissions he's been granted or chores he may perform only with special approval. Skills are things such as a regular weekly allowance or how much TV/computer game or play time they are granted each week. Skills can also be how late the child can stay up on school nights, whether he can take an elective course, be on a sports team, or use the lawn mower unsupervised. It's anything your child can do that you should be monitoring as part of being an involved parent, and anything that can be seen as a reward for meeting goals and expectations.

Certain skills are "growable," meaning they can be increased over time. We call this "adding ranks." For example, when your child is six or seven, getting $1 a week in allowance may be sufficient, so he may choose to take one Skill in Allowance ($1 per week per rank). As he gets older, and financial needs increase (must buy more video games!), he may choose to spend the skill points he gets when leveling up by increasing his rank in Allowance to get them up to $3 per week. The same may work for minutes of television per week, or school night bedtime.

Other skills may have limited growth yet also allow the earning of more experience points. For example, Yard Work. At rank 1, this means the child is allowed to help an adult with yard work for small amounts of eps. At rank 2, the child is allowed to help with yard work, including supervised use of the lawn mower and hedge trimmer. This will allow him to earn higher amounts of eps for the challenges he faces. At rank 3, the child is allowed to perform yard work completely unsupervised, earning even more eps every time he does the job. Of course, as these skills require a certain level of maturity and care, you can make them harder to attain by increasing the skill points required to buy additional ranks, or set a minimum age for choosing them. Once again: Tailor the system to your children!

Suggested Skills, Point Adjustments (Tags)

1-POINT SKILLS

- Bivouac—+1 to Challenge Rolls for cleaning a room (cleaning, solitary). Rankable.
- Scullery—+1 to Challenge Rolls for doing the dishes (cleaning, solitary). Rankable.
- Animal Handling—+1 to Challenge Rolls for pet care (feeding, scooping, washing, walking). Rankable.
- Athletics—+1 to Challenge Rolls for successes in sporting endeavors (practice, games, outdoor play). Rankable.
- Academics—+1 to Challenge Rolls for successes in educational endeavors (tests, spelling bees, weekly homework, reading goals). Rankable.
- Minstrelsy—+1 to Challenge Rolls for successes in performance (plays, chorus, arts, band/instrument practice). Rankable.
- Salary—+$1 per week in allowance (benefit, monetary). Rankable.
- Curfew—+30 minutes to base bedtime.
- Entertainment—+30 minutes a week to allowed TV, video game, computer game time. Rankable.
- Indoor Tool Use (Specific Tool)—Permission to use a specific indoor tool toward confronting higher-exp challenges (washing machine, iron, vacuum, stove, oven, etc.). Two ranks: 1 point allows supervised use, 2 points allow unsupervised use.
- Outdoor Tool Use (Specific Tool)—Permission to use a specific outdoor tool toward confronting higher-exp challenges (lawn mower, hedge trimmers, leaf blower, pressure washer, etc.).

Two ranks: 1 point allows supervised use. 2 points allow unsupervised use.

- Environmentalism—+1 toward Challenge Rolls on environmentally friendly activities (recycling, composting, park cleanups).

2-POINT SKILLS

- Indoor Combat—+1 to Challenge Rolls for all indoor chores, such as room cleaning, laundry, dishes (does not include tool use). Rankable.
- Outdoor Combat—+1 to Challenge Rolls for all outdoor chores such as basic yard work, simple maintenance, or cleanup (does not include tool use). Rankable.
- Cooking—+1 to Challenge Rolls for fixing family meals. Rankable.

3-POINT SKILLS

- Mastery (Challenge)—+5 to Challenge Rolls for a specific challenge for which the character has taken over complete responsibility (per GM's determination). For example, if the character has taken on doing all the family laundry, every day, he could take this skill and be relatively assured of earning high bonus eps every week. Must already have skills for any tools required.
- Driver's License—One rank: Authority to take driver training and obtain a learner's permit. Two ranks: Permission to take the test and get the license. Three ranks: Permission to use the car supervised. Four Ranks: Permission to use the car unsupervised.

Challenges/Experience/Levels

The basic idea of most RPGs is rolling a die to determine success or failure at a challenge based upon your character's abilities, skills, and a luck factor. Success (and sometimes failure) brings experience, and characters will "level up" after accumulating experience—attain a discrete new level of understanding and skill where they may get new points to spend on increasing their skills and attributes.

For our game, the challenges won't be imagined combat with subterranean monsters or solving a puzzle to disarm a trap; rather, these are the chores our kids must do regularly, the tests we want them to study for and perform well on, or the sports, music, or theater events we want them to practice for and perform well at. So, since the characters in our game (our kids) will actually be performing the challenges we give them, we won't have them roll a die beforehand to determine success or failure. Instead, we'll have them roll a die after the job is done, to determine how many experience points they earn from the job they did.

Below are some suggested challenges with base exp values and tags to help determine which skills and bonuses apply. This list is by no means exhaustive, and you should work up your own list based on your kids' habits and talents. What is very important, though, is balancing the exp they earn. The game is set up based on the idea that kids will gain one level every three to four months. For younger kids and those just starting out, that means twelve to sixteen weeks between levels, which are 1,000 exp apart, so kids should be earning in a range of sixty to ninety exp per week. Like any good game master, you need to balance. Too many points coming in, and the kids will get bored with the game; too few and they'll get frustrated. Use your best judgment, and communicate with them to make sure they're getting something out of it.

Suggested Challenges, Base eps Value (Tags)

- Room Cleaning (weekly), 20 eps (indoor, combat, solitary)
- Dishes (weekly), 20 eps (indoor, combat, solitary)
- Yard Work (weekly or per event), 10–25 eps (outdoor, combat, solitary or support)
- Laundry (weekly), 10–20 eps (indoor, combat, solitary or support)
- Pet Care (weekly), 15–25 eps (outdoor, combat, solitary)
- Homework (weekly), 20 eps (indoor, magic, solitary)
- Instrument/Sport Practice (weekly), 10–20 eps (indoor or outdoor, combat, solitary or group)
- Tests (per event), 10–20 eps (magic, indoor, solitary)
- Performances/Games (per event), 15 eps (indoor or outdoor, magic or combat, solitary, group, or support)
- Church Event (per event), 20 eps (indoor or outdoor, spiritual, group or support)
- Clean Out Garage (per event), 50 eps (indoor, combat, solitary or support)
- Prepare a Meal/Cook (per event, weekly), 20 eps (indoor, creative, solitary or support)
- Paint a Room (per event), 50 eps (indoor, combat, solitary or support)
- Participate in a charity event (per event), 30 eps (indoor or outdoor, support or group)

To determine the experience earned for a task, first start by setting a base experience value for the task. Then determine a base Challenge Rating for the task; default should be 11—the statistical average roll for a d20 (actually it's 10.5, but you can't roll a 10.5, so we round). If your child performs the challenge especially well, you can lower the rating (making it easier to earn a bonus), or if he does it poorly, raise the rating (making a negative more likely). Then determine your child's Challenge Roll bonus based on his Level, Attributes, and Skills. Have him roll a d20, add the bonus, and figure out the result below:

ROLL VS. CHALLENGE RATING	RESULT
Natural 20	Critical hit—Base eps +50%
Roll>=CR + 10	Base eps +25%
CR + 10>Roll>=CR + 5	Base eps +10%
CR + 5>=Roll>CR—5	Base exp
CR—5>=Roll>CR -10	Base eps -10%
CR—10>Roll	Base eps -20%
Natural 0	Critical Fumble—Base eps -30%

For example, your child's challenge is to clean his room once a week, earning 20 base eps each time. The challenge "Room Cleaning" is tagged as "indoor, physical, solitary." Your child's character is a second level Elven Ranger, he has a 15 STR, and took the Bivouac skill with 1 rank so far. His Challenge Roll bonus is +2 for being second level, +1 for STR, and +1 for the Bivouac skill, for a total of +4. You feel he did a reasonable job at the cleaning, so you set the CR at 11. He rolls a d20 and gets a 12, and adds a bonus of +4 to get a total 16 versus the CR, or 5 more. Since he got 5 over the CR, he earns a +10 bonus to the eps—HOWEVER, because he is an Elf, he loses 10% eps on indoor activities, leaving him with the base 20 eps.

Yes, this is math. But your kid will do it, and enjoy it, because the risk/reward is awesome, and most kids love working this kind of thing out (and trying to game the system—watch out!).

Eventually, your players will gain enough experience to level up. The chart below sets eps goals per level, and the bonuses they gain, for a span of six to eight years of the game. Again, the goal is to have them level up about once every three to four months. If your kids play longer, expand the list as you deem fit.

LEVEL	EPS	CROLL	ADD SKILL PTS.	ATTRIBUTE PTS.
1	<=999	1	BASE	BASE
2	1,000	2	+1	-
3	2,000	3	+1	-
4	3,000	3	+2	+1
5	4,500	4	+1	-
6	6,000	4	+1	-
7	7,500	4	+2	+1
8	9,000	5	+1	-
9	11,000	5	+1	-
10	13,000	5	+2	+1
11	15,000	5	+1	-
12	17,000	6	+1	-
13	19,000	6	+2	+1
14	21,500	6	+1	-
15	24,000	6	+1	-
16	26,500	6	+2	+1
17	9,000	7	+1	-
18	31,500	7	+1	-
19	34,000	7	+2	+1
20	37,000	7	+1	-
21	40,000	7	+1	-
22	43,000	7	+2	+1
23	46,000	8	+1	-
24	49,000	8	+1	-

Character Sheet

Of course, they'll need a character sheet to track all of this. You'll want to record, well, everything. Indeed, the character sheet will be a living document that gets updated every week, if not every day, sort of a checkbook combined with a personal journal. And while your geeklets may chafe at the paperwork to start, in a while the value of it as a personal record will become obvious. Plus, when you say something about their not mowing the lawn last week, they'll have all the proof they need to refute your spurious claim. Hey, wait a minute. . . .

There's a sample character sheet included in Appendix B at the back of this book. It's pretty basic, but it'll give you an example of what to track. It's a good idea to have extra pages attached as a register of the challenges faced, the eps earned, and the skill and attribute points gained and distributed.

More Resources

Everything above should be plenty to kick things off, especially with a little thought on your part about the challenges and skills you want to use. There are downloadable versions of the charts and character sheets on www.geekdadbook.com, as well as forums where we hope people will share their additions or variations on this game.

A Never-Ending Demolition Derby

The demolition derby has always held a special place in our culture. Not unlike Ultimate Fighting, the spectacle is simple: Two cars enter, one car leaves. Problem is, once someone wins, someone else has to go looking for a new car. Actually, even the winner often has to look for a new car. Kind of a waste of materials, really.

So why not build cars for the demolition derby that are designed with breakaway components that can be reattached or easily replaced after each match? Then the competition becomes a matter of the strategic ablation (knocking pieces off) of your opponent's vehicle, rather than a simple smash-and-bash.

Of course, if we can imagine something like this on a big scale, we ought to be able to build and play it on a small scale.

PROJECT	A NEVER-ENDING DEMOLITION DERBY
CONCEPT	Attach LEGO base plates to the exterior of an R/C car. Build up structures all over it, and then play demolition derby with an opponent. Create rules and a scoring system.
COST	$–$$
DIFFICULTY	⚙ ⚙
DURATION	☼ ☼ – ☼ ☼ ☼
REUSABILITY	◍ ◍ ◍
TOOLS & MATERIALS	2 or more R/C cars, LEGO bricks and plates, peel-'n'-stick Velcro or foam tape or a hot-glue gun, foam pieces (optional, as needed)

To start, you have to select your R/C cars very carefully. It's going to be much easier to attach the LEGO plates to the outside if you have plenty of flat surfaces rather than round ones to work with. With flat surfaces, all you really need to do is select plates that roughly fit on the roof, hood, trunk, and sides. It doesn't matter if you use a few larger plates, or more and smaller plates—the goal is the same: Make as much of the exterior surface of your vehicle that is able to have blocks attached to it.

WARNING: You may be about to permanently affix the LEGO plates to the sides of your R/C car. Do not purchase or use R/C cars that you, or someone else in your family, are attached to from an aesthetic or sentimental standpoint. The cars will be irreparably altered by this project, as could your relationship with loved ones if you hack their toys without permission.

Time for the sticky-sticky. As I suggested in the project summary above, you have a number of options for attaching the LEGO plates to the cars. Simplest is probably the old crafting standby, the hot-glue gun. A couple of globs per plate and the job is done, though you may need to give them some curing time to make sure the glue has dried.

There is also the two-sided foamy tape used for its ability to stick things to walls, and cursed for then taking layers of drywall paper with it when removed. This tape has the advantage of allowing you to build up layers to help deal with contours on the cars' surfaces as well, though you need to make sure you've got your placement right the first time.

Another cool alternative is to use a roll of two-sided sticky Velcro tape (perhaps in combination with the foam tape). This will give you the flexibility of being able to remove and reaffix plates, perhaps in mid-derby, allowing for some interesting rule options. Or, for maximum remove/reattach action, the stick-on/pull-to-remove sticky strips used for picture hooks these days will do an admirable job as well.

I mentioned regular craft store squishy foam as an optional material. This would be useful for fine-tuning the attachment of plates if it is trimmed into custom shapes to help fill in dips and rises in your cars' shells. Entirely up to you to use or ignore.

With the plates attached, you've done the "hard" work (and really, it's not that hard, is it?). Now it's time for the creativity:

- Build a seat on the roof for your minifig, with a simulated cage around it so he looks like the Mohawk guy from *Road Warrior.*
- Build some kind of ramming construct on the front of your car. Or a scoop to try and flip your opponents.
- Put bricks on the sides to act as armor.

Once you and your kid have your wicked-looking combat cars assembled, it's time to SMASH!

But there's more you can do besides just bashing into each other! Why not turn it into a game? Decide what kind of game you want your demolition derby to be, and build brick structures on the base plates accordingly. Here are some ideas:

- Each player gets a certain number of bricks to plug on to his car. Hold a timed battle (say, two minutes), and the car with the most remaining bricks at the end wins the round. Hold multiple rounds with multiple challengers to create a tournament.

- Each player has a group of specifically colored bricks that he plugs into the outside of his car. Then he can build up structures around those bricks to protect the special bricks. Hold a timed battle, and each colored brick that's knocked off earns the opponent a point. Score points to win games, win games to win a set, win sets to take the match!

- Each player has a minifig "driver." The driver is attached on the roof (hopefully, using a LEGO chair of some sort), and then structures are built on the sides and around the minifig to protect it. Hold a battle, and the last minifig still attached wins.

For all the game scenarios above, if you use the Velcro idea above, allow time-outs where plates/structures can be swapped out to rebuild a car mid-battle.

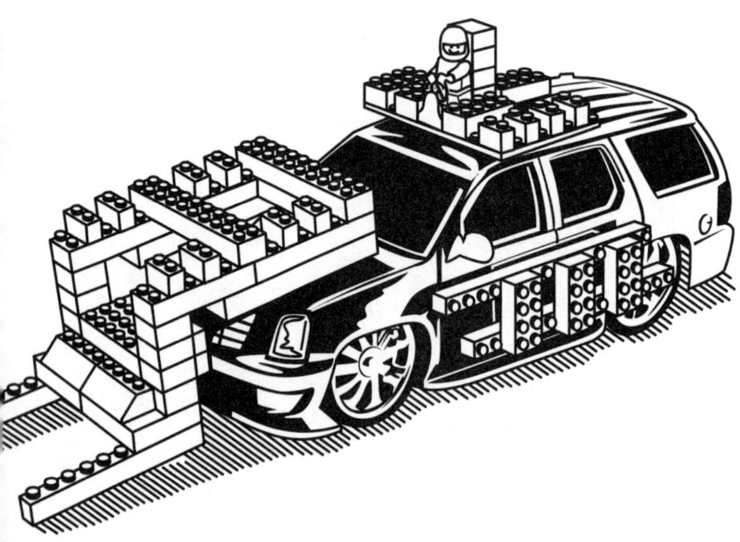

An Even Geekier Idea!

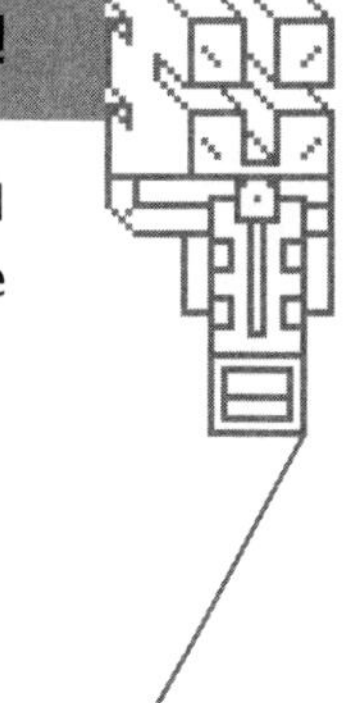

To add to the derby sensibility, you can build an arena for all this model carnage. If you have some spare 2-by-4s, lay them out in a useful shape (octagons are cool), and use duct tape and a staple gun to temporarily connect the ends. For an easier and potentially cheaper approach (especially if you do the Slip 'N Slide project on page 113 in this book), use pool noodles as the borders of your arena, and try duct tape and rubber bands to loosely tie the ends together.

AND A NOTE ABOUT CLEANUP: One potential time waster with this project is having to clean up all the bricks between battles. There is an easy way around this problem. If you're like most GeekDads, one key feature of your workshop is a Shop-Vac. Make sure it's clean and empty at the start, and then just vacuum up all the cast-off pieces between battles, and dump them back in the construction pool when it's time to build some more. Easy!

Okay, GeekDads and Kids—time for some vehicular carnage!

GEEKY ACTIVITIES FOR THE GREAT OUTDOORS

See the World from the Sky

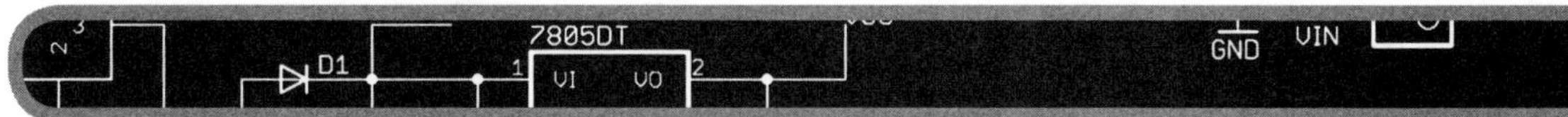

In the last couple of years, there has been a spate of stories about enterprising students and other private citizens building and sending amazing balloons, with cameras and other instruments attached to them, miles into the sky. MIT student Oliver Yeh did it with $150 worth of materials (including a secondhand camera and a Styrofoam cooler) in September 2009. These packages then return with incredible pictures, providing yet more corroboration of the curvature and beauty of our planet. Such projects have to overcome some interesting technical challenges, such as GPS tracking and the necessity to keep equipment working in very cold, low-pressure conditions. Really, they are a kind of UAV (Unmanned Aerial Vehicle).

While launching a balloon several miles into the sky is a tremendously exciting project, it is not the kind of thing you can throw together with your kid over a weekend. This project, though, is designed to help you have the same kind of fun at lower altitudes and for significantly less expense (of both money and time).

PROJECT	SEE THE WORLD FROM THE SKY
CONCEPT	Make your own (relatively) high-altitude video package.
COST	$$–$$$
DIFFICULTY	⚙ – ⚙⚙
DURATION	☼–☼☼
REUSABILITY	⊕⊕
TOOLS & MATERIALS	2–3 helium balloon party kits from Walmart or Target, a dozen balloons, zip ties, duct tape, kite string, Styrofoam, video camera (Flip or similar)

A little research on the Internet tells us that, based on the lifting capacity of helium, a group of balloons with a volume equivalent to a 4-foot-diameter sphere (around 12.5 cubic feet) should be able to lift about one pound of payload. So, to pull this project off, we have to keep the mass of our entire build under that limit and/or fudge the number of balloons (and thus total helium volume) so we can get a little extra lift. If we can inflate the "normal" balloon that comes with the kind of party kit you can buy anywhere to about one foot in diameter, then we'll need about 16 balloons to get the kind of lift we want (each one-foot balloon would hold about 0.75 cubic feet of helium).

The overall design concept is this: Build a "column" of helium-inflated balloons, not unlike what you see at fancy parties, hook a camera to it, and let 'em fly while keeping them tethered to the ground (you) with kite string. The camera itself needs to be set securely into some kind of padded open-face enclosure so that it won't be jarred too severely upon a rough landing (though we're hoping for a very controlled descent). And that enclosure has to be connected easily to the kite string so that the camera doesn't go flying a lot farther than you want it to.

The balloon method I suggest using for this project is not the only way possible, of course. A kite could work, and could potentially be somewhat less expensive and require purchasing fewer materials. However, most inexpensive kites are not as inherently stable as balloons. Balloons want to go up and stay up, and it's you keeping them from doing so, whereas a kite, unless it has caught the wind just right, wants to come down and crash into the ground. And because for a kite to work properly, it must be a windy day, it won't keep an attached video camera as steady as will a placid bundle of balloons on a windless day (and please, let me stress that you want to do this project on as windless a day as possible).

It is at this point that the pro-kite community will argue that there are kites that can be very stable when aloft. This I do not contend. However, I'll point out that such kites will cost at least as much as the helium we're using, so we'll say touché, and move on.

A Flip digital camera weighs just slightly more than five ounces, making it an excellent choice for the technical package. Plus, a lot of geeks already have one, so it may not be an extra expense for you for this project. If you have a different camcorder, check the weight. Anything much more sophisticated than the flash memory-based models available right now (Flip, Kodak, and similar) will likely weigh too much.

Of course, sending a valuable piece of technology up into the sky, tied to a bunch of balloons, may get you an odd look from your spouse, but we have no good suggestions here to help you with that, other than telling her or him it's for science!

BUILDING THE CAMERA PACKAGE

STEP 1: Because you'll want to get your video camera back, you'll need to tether your craft, and even if the kite community is still smarting from the logical thrashing I just gave them, I do suggest using kite string since, really, it's made for this sort of thing. One reference I found online cites a weight-to-length ratio for kite string as 1 pound for 8,700 feet of string. If we restrict ourselves to a commonly available 500-foot line, that adds only 0.9 ounce to the package when it reaches full height.

STEP 2: Next, we need a cockpit for your camera (no sniggering). My choice was a piece of Styrofoam I picked up at my local hobby shop. It was actually purposed for carving out mountains for model train landscapes, but the size and strength were right to work with, and the density was great for cutting with a chop saw. Foam packing material that comes in standard toy or equipment packaging may work as well, though it's usually less dense and thus harder to cut cleanly. On the other hand, that means it'll weigh less, so there is a fair trade-off.

STEP 3: We don't need the piece of Styrofoam to be huge, so just cut it down with a chop saw or serrated knife to about double the size of the camera itself. Then trace the camera's outline on one side (with a little fat), and use a mat knife to trim out a cavity into which the camera can be set.

STEP 4: To attach the anchor points to the camera, get your duct tape ready. Lay two zip ties (unzipped) laterally across the top of the foam block and then tack them down with a small square of tape. Do the same on the bottom. If you are an engineer like me, and want to feel you're accounting for stress lines and such, have each zip tie in a pair oriented in opposite directions so that the points and heads are reversed. This is probably completely unnecessary, but don't tell your kids that; rather, impress upon them the vital importance of this key design factor in ensuring stability and structural integrity. If they ask "Why?" just shake your head and tell them, "Someday, you'll understand."

STEP 5: With the zip ties tacked on at the top and bottom, take the duct tape and, starting from the back side of your package, make two complete passes down, under, up, and over, pressing the tape into the camera cavity each time you go around. Finish the tape on the back where you started. Now repeat this, going horizontally around the package, from back, around the front, to the back, and around again, pressing the tape into the cavity each time. Your finished product should look something like this:

With all this done, you're about ready to fly. All you need is your anti-gravity technology!

BUILDING THE BALLOON COLUMN

To tie the balloons, you can use the classic thin iridescent ribbon that comes with many party tanks of helium and balloons. While it's cheap stuff, it is very lightweight and has, as anyone who has tried to open a present by pulling the ribbon apart knows, tremendous tensile strength!

Building the balloon column is truly a partnering activity. One person should be inflating and initially tying each balloon while the other is assembling the column. The work flow should be as follows:

1. Partner #1 (P1) inflates a balloon to about 90 percent of capacity and ties off the end with a single knot. Then hands it to P2.

2. P1 inflates a second balloon to 90 percent capacity, ties it off, and hands it to P2.

3. P2 takes the ends of each of the two balloons and ties them, using the ends of the balloons like string, double-knotting them together.

4. Repeat the first three steps.

5. P1 and P2 each take a tied pair of balloons. Hold them together with the tied centers crossing each other at a 90-degree angle. Twist the balloons around each other a couple times so that the tied centers become intertwined (rather like what you'll see balloon-animal artists do to build their creations). Done properly, you should now have four balloons in a flowerlike shape.

6. Take one end of the ribbon and knot it around the intertwined middle so the balloon flower now has a string. Do not cut or trim the ribbon—leave it in its packaging so you can pull more from it while it's connected to the balloons.

7. Repeat the first five steps to make another quad of intertwined balloons.

8. Take the ribbon that's attached to the first quad of balloons, move down it about 6 inches, and loop the 6 inches of ribbon around the intertwined middle of your next quad of balloons, going up, around, down, under, in, and out, almost as if you're weaving the ribbon around each axis at the middle of the balloon. Finish up so that the ribbon can continue "down" to the next quad to be added.

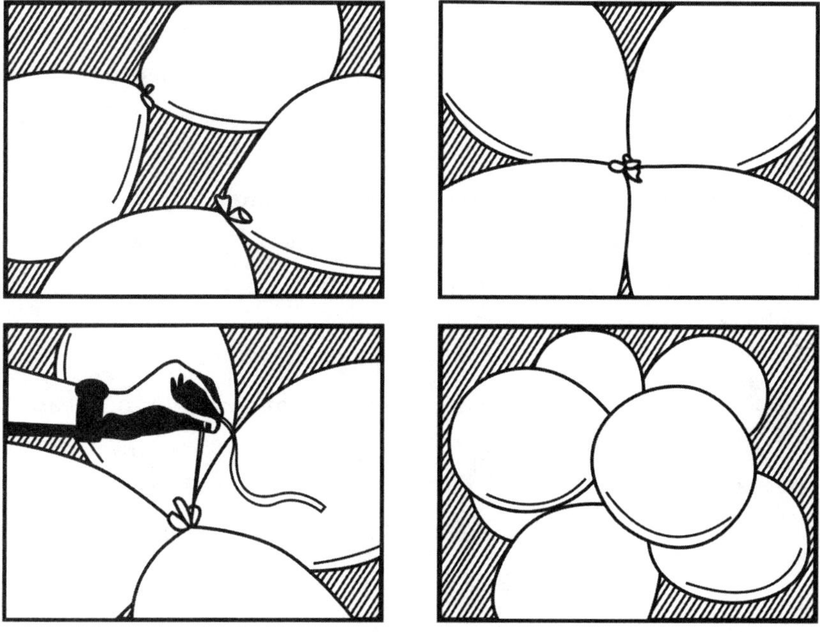

9. Repeat the previous two steps until you have at least four quads of balloons tied together into the column.

How many balloons you'll need will depend on the size of the balloons you get in your party kit. Usually they'll inflate to a pear shape about 10 inches across, and 14 inches long. Once you have sixteen of these, in four quads, tied onto your ribbon, you can test the lifting capacity by running the ribbon through the wrist strap of your camera (which is the heaviest part of your build) and seeing if the camera lifts off. Keep in mind that you want it to lift pretty sharply—we don't want neutral buoyancy here, we want to have a pretty strong lift. So if the balloons lift the camera, but weakly, add another quad of balloons to be sure.

When you have your balloons done, cut the ribbon about 18 inches below the bottom quad of balloons, and use all your Boy Scout or Wikipedia know-how to tie a knot in the end that creates a nonslipping loop. Pass the ends of the zip ties on the top of your camera cockpit through this loop, then zip them together so the balloons are securely attached to the package. Play with the zip ties so that, when the balloons pull upward, the package is oriented as straightly vertically as possible, for better filming.

Get your kite string and tie the same nonslip loop knot in the end of it. Do the same thing with the zip ties at the bottom of your package to secure the line.

Start the countdown!

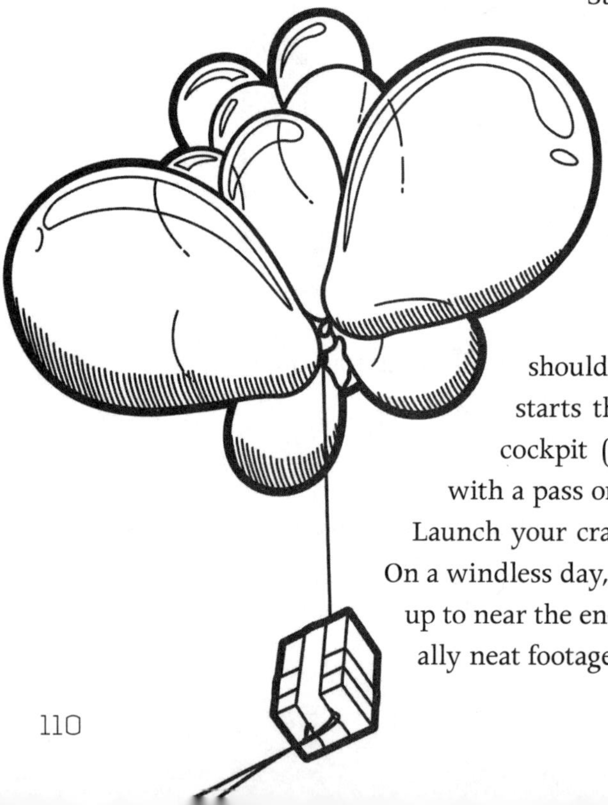

Now you can head out to a nearby field, park, or other open area where you can get some clearance around you, just in case. Take your camera (fresh batteries, cleared memory, please!) and the roll of duct tape with you.

When you're ready, one person should hold the balloons while the other starts the camera recording, sets it into the cockpit (lens facing outward!), and tapes it with a pass or two of duct tape.

Launch your craft, and let the balloons take it away. On a windless day, you should be able to get the camera up to near the end of your kite string and get some really neat footage. Then, just like pulling in a kite, rewind your tether to bring the craft back to Earth. Go home, hook your camera up to your big-screen TV, and get a feel for what it's like to be a bird (and find out which of your neighbors needs to clean their pool)!

An Even Geekier Idea!

For a bit more money, this project can turn into an even more amazing experience. You can purchase a wireless, battery-powered video camera (often sold as "surveillance" cameras) to put into the balloon package instead. Pick up a pair of video-projecting glasses, and watch the footage real-time!

Best Slip 'N Slide Ever

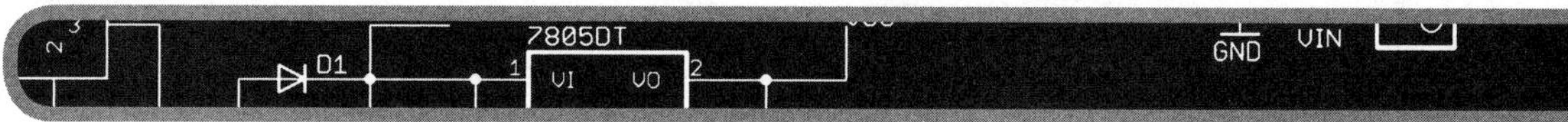

When I was a kid, I remember building a homemade Slip 'N Slide with my friends to have some outdoor fun on a hot summer day. We'd usually cut up a number of black garbage bags and try to overlap them to create a good run. Then turn a sprinkler or two on them, and get busy.

These days, mass-produced Slip 'N Slide–type things are available at any big-box store for around $30. They're big, bright, and even imaginative. Heck, you can drop a couple hundred bucks and get giant inflatable water slides that will fill up your whole yard.

What I've found over a few years with my kids and their friends is that the quality of construction usually makes these slides a one- or two-use product. And while I said they were imaginative, they're usually not that big, since they're designed for a mass market of people who won't all have the yard space for a larger slide. So I started to wonder if there wasn't something that could be done at home, in the DIY spirit of using garbage bags like I did as a kid, but a bit more durable and, you know, BIGGER. And what I came up with is easy to build, hugely fun to play with, durable, and simple to take apart and store for significant reuse.

PROJECT	BEST SLIP 'N SLIDE EVER
CONCEPT	Build your own awesome Slip 'N Slide.
COST	$$–$$$
DIFFICULTY	⚙
DURATION	☼☼
REUSABILITY	◍◍◍
TOOLS & MATERIALS	Heavy sheet plastic, pool "noodles," peel-and-stick Velcro, some kind of sprinkler or sprinkler hose

This project may have the claim to fame of being the largest-scale but easiest-to-build in the book. We're putting together the basic concept of the Slip 'N Slide (SNS) using durable over-the-counter materials. All you need is a $30 roll of heavy plastic, ten $2 pool noodles, a $10 sprinkler hose, and a couple rolls of peel-and-stick Velcro (about $7 a roll).

So what is an SNS at its core? It's simply a long expanse of material that gets slippery when wet. It should have some kind of guides or berms on the sides to keep sliders from slipping off while traveling down its length. And it needs a water source.

1. To start, take your roll of heavy sheet plastic and lay it out on your yard or other assembly site. We tried some 6-milliliter plastic, 6 feet wide by 50 feet long for our sample slide, since it gave a nice width of sliding surface, and the length fit across our front yard. But, depending on your location, you may want a smaller or bigger (yeah!) slide. Figure out which side is the top (it's a completely arbitrary decision since both sides are the same, but you have to pick one and stick with it), and place it facedown.

2. Lay the noodles around the perimeter of your plastic. You can leave a foot or so between each noodle. The standard length of

a pool noodle is about 5 feet, so for our 50-foot long slide, we used eight noodles per side with about a foot of spacing, give or take, and then one noodle at either end.

3. Next, starting at one end, take a noodle and lay it on the plastic a few inches in from the outside edge. Pull the plastic over the noodle as if you're going to wrap it up, and get enough overlap so about an inch of the plastic from the edge touches the plastic on the other side of the noodle toward the middle. This is where you'll be sticking the Velcro.

4. Attach a 2-inch strip of the Velcro to the plastic at each end and in the middle of each noodle so that the plastic wraps over and under the noodle and is attached back to itself. Do this for all the noodles until you have a berm all the way around the perimeter of your slide.

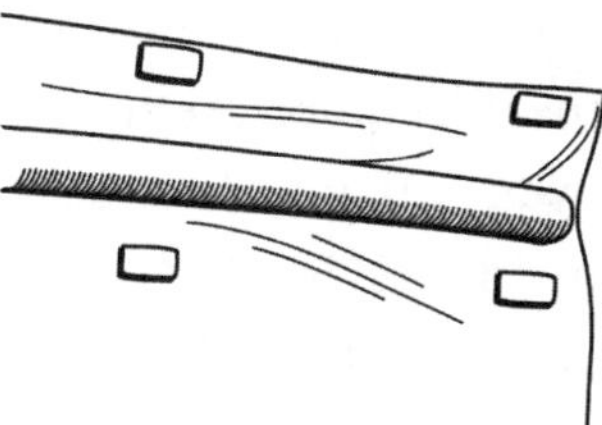

5. Once you're done, you have the underside of your slide. Flip it over, and you should have what looks a little like a very long, very narrow emergency slide from an airliner. Or a really cool waterslide.

6. Last thing we need is the water source. If you're keeping it simple, just make sure you have a little slope and start running a hose at the top of the slide at the higher end (where you'll start your slides from). Or if you have one or more lawn sprinklers, use those. For a little more money, pick up a 50-

foot sprinkler hose and (if you also got the extra roll of Velcro) affix it to the side of the slide down one of the berms. Use your regular hose to feed water into it and you've got a perfect shower down your slide.

IMPORTANT TIP FOR A FLAT YARD: The best placement for an SNS is on a gentle downhill slope that peters out at the end to flat, but not all of us are lucky enough to have the perfect sliding real estate. If you're building your SNS on a flat expanse of lawn, an added feature could be of use. Get a piece of rope about 6 or 8 feet long. Tie each end to a short piece of wooden dowel or a plastic handle like the ones that come with car window squeegees or toilet plungers. Make sure you have good knots, and perhaps wrap it all up in duct tape as well. You now have a towline. Position your child to sit at the starter end of the slide, either in a crisscross applesauce position or on his or her front or back, and have someone as big or bigger pull him while running down the slide. Once the initial friction is overcome, it's not very hard to build up a bit of speed down the slide. Just make sure to have them let go before the end of the slide.

Now just wait for a warm day, collect the neighborhood kids, and become the best house on the block! Oh, and in case you hadn't noticed, with all that Velcro, this thing is really easy to disassemble and fold away for another day.

Fireflies for Every Season

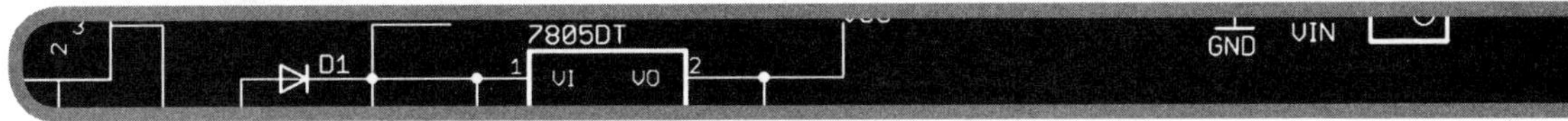

An enduring memory from the childhood of many people who grew up in rural or semirural areas is the sight of fireflies on a warm summer night. Watching the clouds of sparkling lights dancing around is a magical experience. But for people in urban areas, fireflies are harder to come by. Even if you do live in a firefly-friendly area, the summer lasts only so long. With this project, you can enjoy the magic of fireflies anywhere, at any time of year!

PROJECT	FIREFLIES FOR EVERY SEASON
CONCEPT	Build simple electronic fireflies to play with.
COST	$$
DIFFICULTY	
DURATION	–
REUSABILITY	
TOOLS & MATERIALS	CR2032 3v batteries, 5mm yellow LEDs, fabric or electrical tape

[This project was developed by GeekDad writer Dan Olson.]

We are busy parents, so sometimes even if we have the inspiration for a fun project to do with our kids, it takes some time to put our idea into action. I was inspired to make a cheap solar light based on an www.instructables.com post involving deconstructing solar garden lights and reinstalling them in Mason jars. My inspiration held out long enough for me to search fruitlessly for some cheap solar lights, but not long enough to make anything.

Fortunately for me, the Internet provided the answer rather quickly (via an especially good geeky project site called Evil Mad Scientist Laboratories—www.evilmadscientist.com). And I was finally able to make my dream happen!

For this project, purchase the following from your local electronics store (or order them online):

- 12 (or more) CR2032 3v batteries
- A bag of 25 diffused 5mm yellow LEDs
- A twenty-five-cent roll of electrical tape

Total cost was under $20. Each firefly costs about $1.50.

The build process is very, very easy:

1. Unwrap the battery.

2. Slide one leg (lead) of the LED onto each side of the battery. It should light up. If it won't light up, flip the LED around for a quick polarity lesson. The longer leg is the positive cathode.

3. A little tape around the battery covering keeps it lit.

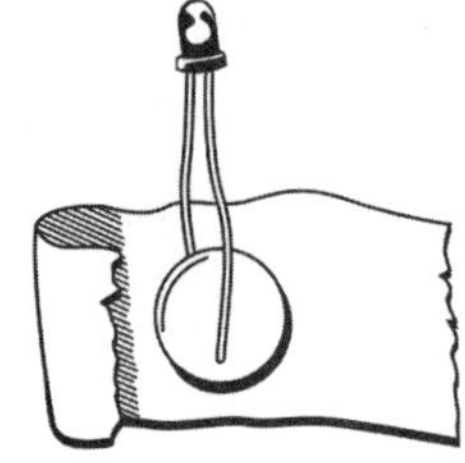

At my house, proper presentation is the key to early adoption. You can't be too excited about a new project or the kids will go back to the couch. But if you casually toss a lit firefly on the table and instruct them to stay out of the bag with the rest of the materials, they'll have ten of them glowing before you come back from the restroom.

This project teaches some basic lessons about electronics as well. When I made these with my own kids, they started testing multiple LEDs on the same battery. Luck (and some bent LED leads) taught them that switching the power on and off is the result of the proper contact being made between LED leads and the appropriate positive or negative side of the battery.

Once you've created your all-season fireflies together, send your kids out into the backyard and enjoy the fun as they run around in the dark holding the fireflies aloft.

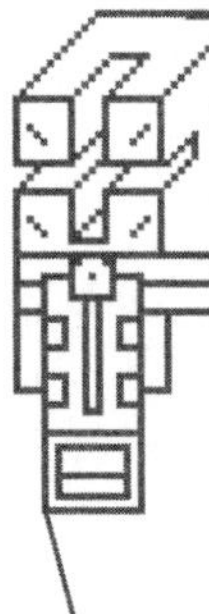

An Even Geekier Idea!

- Grab some multi-LEDs and switch the polarity. They'll switch colors.
- Make "throwies" by adding a rare-earth magnet to your firefly, taped to the battery. Then walk around at dusk tossing them at road signs and light poles.
- Put a firefly in an empty wide-mouth drink bottle and make a garden lamp. No glass jars are required.
- Tape your fireflies to cheap balsa-wood gliders and toss them around the backyard.

This project is painfully simple, requiring almost no skill and very little cash to achieve an experience to spark the imagination. You just need inspiration to pull the project off. But inspiration can go a long way.

Video Games That Come to Life

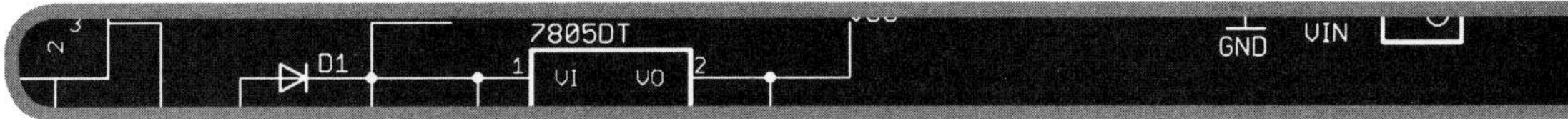

We all have video game machines in our homes. And the games—adventures and shooters and platformers and simulations—are wonderful sources for our kids to learn how to meet challenges, solve puzzles, think logically, formulate critical observations, engage in team play, and even understand basic science and math. The games can be great ways to learn the nuance of sports and military strategy. But even if you have Wii Fit, playing video games isn't a true substitute for going outside and exercising, not to mention playing with friends in the fresh air.

So, as the GeekDads who love to spend hours tethered to a game console that we are, how do we encourage our kids to turn off the machines, get outside, and play the way people did before the Atari 2600 and rampant childhood obesity came along (not that I'm linking the two . . .)? Here's one good idea: Make the games they play outside versions of the games they play inside.

PROJECT	VIDEO GAMES THAT COME TO LIFE
CONCEPT	Make outdoor games more like video games to get kids more interested in playing outside with their friends.
COST	$-$$
DIFFICULTY	⚙
DURATION	⏱
REUSABILITY	◍◍◍◍
TOOLS & MATERIALS	Flash cards, tissue paper, beach balls, brooms, water guns, embroidery hoops

The easiest video games to re-create outside, of course, are sports video games. Why tell your kids to go out and play some Wiffle ball or flag football when you can get them excited about trying MLB2k9 Home Edition or Madden Backyard? The goal is to play to their imaginations. Plain old baseball or football is boring, but add the concept of the video game to it, and then it gets interesting.

What does that entail? Good question! What is it that makes the games cool? Usually it's a matter of playing your favorite teams and players, working the strategies, and maybe the career mode parts of the game that let you act as a team owner, trading players and building your franchise. So why not put some of that into the outdoor games? Can you imagine setting up the rules for picking players and teams? Maybe you set up football play or baseball pitch cards that add an element of the video game play to the yard game. For example, each team in a football game gets to draw a number of predefined play cards and use them to pick the plays they run, offensively and defensively each down.

Or use proper rosters for each side to pick from, so that the kids can play a certain player with special skills pulled from their current game stats. And use the games to help teach the kids about the strategies of the games—what are the right defensive plays in football to react to a given offensive pattern, or what's the best time to bunt or sacrifice in baseball? Just relate everything back to the video game so that each way of playing the sport becomes a reinforcement for playing the other. If the kids like playing Wii Sports Golf, then play a golf game on your lawn with Ping-Pong balls, but incorporate real golf rules so the kids actually learn about the game and why the things they see in the video game are done that way.

If your kids and their friends aren't exactly into the sport simulation games—that much reality can be boring—what about the fun variations available? The Super Mario and Backyard brands of sports video games add all sorts of fun arcade-style features to the traditional sports games, and you can, too. Try playing Wiffle baseball, but make up a set of game cards allowing all sorts of wacky tweaks that the kids can play for themselves or against each other. For example, "All tied up: Next batter must bat one-handed" or "Superspeed: Next base runner may run home from second base to score." Then for every run scored, a team may pick a card and use it when they deem fit.

Or substitute other items for the equipment in the games. Beach balls and tennis rackets for baseball, brooms and a Wiffle ball for field hockey, or any other variation that makes the game zanier and more yard-friendly is great.

TRY THIS GAME: ULTIMATE OUTDOOR OBSTACLE COURSE

But there's more. What would make a better outdoor obstacle course type of challenge than Sonic the Hedgehog? If you have a big enough backyard or can use multiple front yards to set up a course, you're all set.

1. Get a bunch of items—I like wooden embroidery hoops because they're like the rings in Sonic, and they go for less than

a buck apiece at your local crafts store—and set them along a path.

2. Mix it up with some safe and sane obstacles like paint buckets to jump over, or interesting kids' climbing structures to incorporate (got any large appliance boxes or kiddie pools?).

3. Each kid gets to run the course with you on the stopwatch. (I've found that kids have a passion for being timed, no matter what the task.)

4. Scores are tallied as a combination of the number of items picked up and the speed with which the course is finished—just like Sonic collecting rings while speeding through his levels.

If your kids are more into the cops-and-robbers shoot 'em up, just about any of the FPS or combat games could be ported to outdoor play. What kid wouldn't want to be Master Chief, after all? The biggest challenge with such games, though, is always identifying who actually got hit. How many such games in our childhoods ended in shouting matches of "I GOT YOU" versus "NO YOU DIDN'T!"?

So, one way to solve this in a proper warm summertime game would be to wage a squirt-gun battle with a twist. Kids love to play, but they always want to know who actually won, and they hate subjectivity in any contest, so why not make it easy to know who got hit?

Yes, I know, it's a squirt-gun battle. Pretty easy to know when you've got hit. But how about this: What if it was more about squirt-gun shooting skill? What if we could make a squirt-gun battle as easy to score as a laser-tag battle, without needing to buy all the equipment? All you need are safety pins and Kleenex.

TRY THIS GAME: AQUA TAG

1. Cut out 4-inch squares of Kleenex (colored is better, because it'll stand out more when it gets wet).

2. Pin these targets to the kids' clothes.

3. Each kid gets one fill-up of ammo.

4. Set a time limit, and whoever has the most (or any) dry targets at the end wins the round.

5. If you go multiple rounds, and wet clothes would soak the targets before they get shot, try backing the Kleenex with white

printer paper or (even better) wax/parchment paper to keep the target dry.

A great alternate to the Kleenex is Alka-Seltzer. Yes, I said Alka-Seltzer. If you've got some Alka-Seltzer tablets, a drill, some string, and lots of water, you've got all you need for a fun and cool-looking variation.

TRY THIS GAME: FOAM TAG

1. Drill a small hole through the center of an Alka-Seltzer tablet (it works if you're careful and have the drill bit running when it touches the tablet—too much weight on the tablet will break it).

2. Thread a string through the hole. Make sure the string is long enough to tie around a person's neck so the Alka-Seltzer can be worn as a necklace.

3. The object of the game is to get the other person's Alka-Seltzer wet. As the tablet reacts to the water, it will start foaming and eventually fall off the string.

4. The last person with an Alka-Seltzer necklace is the winner. As an alternative, you can divide into teams instead of every-person-for-himself. (Special thanks to GeekDad Russ Neumeier for that idea.)

But do you notice the one key thing about both these suggestions? Yeah, they include YOU going out, organizing, and sometimes refereeing the play. While it's great to physically play with our kids, sometimes it's just as important to teach them how to play and to help run that play so that everyone gets as much fun out of it as possible. Just like the video games, the play needs to have structure. I think you'll find kids a lot more eager to go out and play when there's a clear purpose and well-defined rules than when you just tell them to "go outside and play."

Fly a Kite at Night

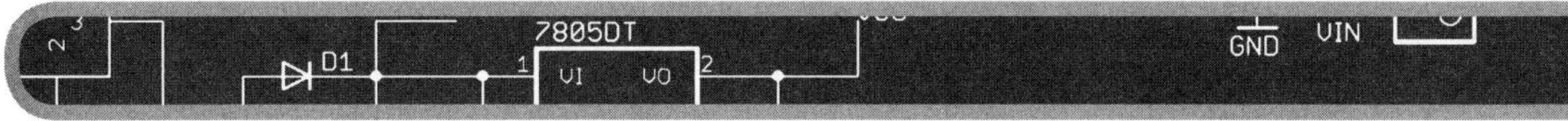

More than two millennia before Bernoulli started learning math, people were building and flying kites. The simplicity of this most basic form of flying machine has entranced both young and old. Anyone with a couple of sticks, some fabric or paper, and some string can spend an afternoon connected to something they have created, and watch it soar with the birds.

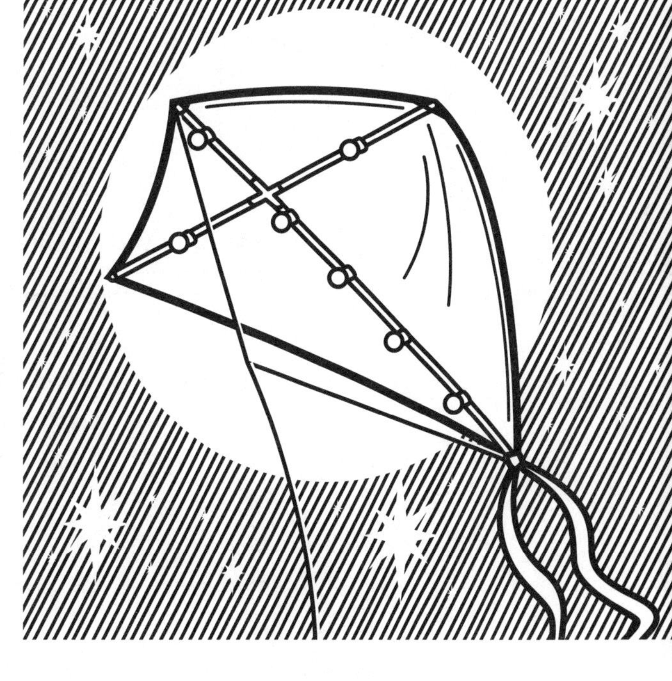

And because of the science behind kites (which we understand better in modern times), many geeks over the years have turned their obsessive gazes to them. From stunt and fighting kites, to the kinds you can strap yourself into and actually fly in, there's a lot of geeky fun to be had playing with kites.

One thing you'll notice about most kite flying, though—it's done during the daytime. This makes sense, since the fun of kites is usually in the watching, especially when it comes to the pretty dragon or box kites. And when you're flying kites with your kids, a big part of the fun is seeing how high they can get (the kites, not the kids). Obviously *seeing* is the key word there.

But we are GeekDads! We won't let something silly, like a lack of daylight, stop us from sharing the joy of kite flying with our kids. Indeed, it seems like exactly the kind of challenge for us to overcome and turn into a cool activity, which is what this next project is all about.

PROJECT	FLY A KITE AT NIGHT
CONCEPT	Attach lights to your kite, and fly at night.
COST	$–$$
DIFFICULTY	
DURATION	
REUSABILITY	
TOOLS & MATERIALS	A kite, lights (LEDs or other), some means of attaching the lights to the kite (tape, glue, zip ties, magnets)

This is probably the easiest project in the book, but it is also one with the potentially biggest wow factor, especially with younger kids. Which means it is maximum fun!

STEP 1: Get a kite. You may already have one of these (many geeks and even some non-geeks do). The kite may be as simple as the classic crossbar with a rhombus of fabric stretched across it, or as complex as a fancy box kite. For our geeky purposes, we picked a *Star Wars* kite!

STEP 2: Get some lights to attach to your kite. You have a lot of options at this stage. The biggest consideration in making your pick is weight versus brightness. Any lights using bigger than AAA batteries are likely to weigh your kite down too much and prevent take-off (although, if you have a large kite and reasonably strong winds, you might be able to get away with a readily available string of battery-powered Christmas lights). In our research, these were some good choices:

- Peel-n-Stick Magnetic LEDs from www.thinkgeek.com (the only downside—they're sort of single-use)
- Mathmos Wind Lights—Micro-turbine LEDs, also from www.thinkgeek.com (more expensive but reuseable)
- Various Arduino-related LEDs from www.makershed.com (geekier and programmable for cool blinky-blinky possibilities)

STEP 3: Attach your lights to your kite. Depending on the lights you choose, you'll need to pick an appropriate means of attachment. For example, the Wind Lights could easily be attached to crossbeams on your kite, using the small zip ties made for cord management. The magnetic LEDs could be affixed to the fabric by applying a ferrous backing to the other side (though there is a potential for slippage or getting flapped off. Two-sided tape might be better). And the Arduino lights could hang off the crossbars with wires, or even be sewn onto a fabric kite.

STEP 4: Go fly! The challenge here is to find the right mix of conditions and location. You'll need a late evening or nighttime with enough of a breeze to fly your kite, and a place to fly that's open enough so you won't have to worry about not being able to see power lines or trees in the dark. If you've got all that, then you're good to go—so long as you're also ready to be visited by Project Blue Book once the UFO sightings start pouring in.

Build an Outdoor Movie Theater

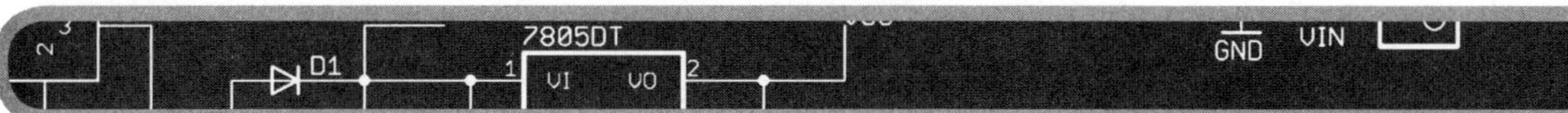

One day, I was looking through a catalog that had come in the mail. It featured everything from a ladder to help an arthritic dog climb onto your bed, to a $6,000 coyote skin throw rug. All pretty expensive, pretty useless stuff. And then I found something that wasn't so useless: an Outdoor Home Theater System.

I thought about it. Lazy summer evenings, sitting back and sipping a cool beverage while the kids ran around catching fireflies and watching a Hollywood blockbuster on a very big screen. These are the moments when vivid, lifetime memories are made. But then I saw the price tag: $3,499 for a projector/DVD combo, two speakers, and a 12-by-6-foot screen. The dream of outdoor movie-watching began to make a hasty retreat.

But before throwing in the towel, I thought I'd explore the idea a little further. What about buying the components on my own? Surely there would be some savings. Sure enough, the catalog listed each component's maker. So I did a quick Google search and found a buy-it-yourself price of $218 for the speakers, $900 for the combo DVD player/projector, and $1,149 for the collapsible screen. It was still more than two grand. For that price, I could practically take the family to Sundance and watch a week of movies.

It was time to get creative.

PROJECT	BUILD AN OUTDOOR MOVIE THEATER
CONCEPT	Make an outdoor home theater setup for a fraction of the cost of retail, and have a blast doing it.
COST	$$$$
DIFFICULTY	⚙⚙⚙
DURATION	☼☼–☼☼☼
REUSABILITY	◍◍◍◍
TOOLS & MATERIALS	DVD player, LCD projector, powered speakers, five-gallon buckets, grout or concrete, PVC cement, eye bolts, anchor stakes, 10-foot sections of 1.5-inch PVC pipe, 10-foot section of 2-inch PVC pipe, 14-inch zip ties, 100-foot nylon rope, 1.5-inch 90-degree fittings, 1.5-inch cross fittings, 1.5-inch tee fittings, piece of blackout cloth cut to the desired size for the screen, grommets, and a setting tool

[This project was originally developed for GeekDad.com by Dave Banks.]

For this project, you've got to get creative with finding the audio and visual items. Many people will already have a portable DVD player, either as a stand-alone unit or as part of a personal or work laptop. A lot of people already have access to LCD projectors as well, usually used for business presentations. For speakers, you'll want something that has its own power source, to amplify the movie sound enough to overcome the ambient noise of an outdoor social gathering.

With the audio and visual requirements resolved, the next step is finding a simple solution for the most expensive element of the equation: the screen. The easiest and cheapest would be a large-enough expanse of lightly painted wall without much texture. That will solve the issue for about 10 percent of homeowners, but for everyone with other colors on our exterior walls, or with siding that would defy a smoothly projected image, something else is needed.

What we need is a screen that is not only temporary but lightweight enough to be portable and storable. The idea we came up with was a screen constructed of blackout cloth stretched over a 1.5-inch PVC frame. Blackout cloth was a great solution, offering nearly perfect color and texture for viewing movies.

BUILDING THE FRAME

You'll want to plan ahead and work with your kid on the frame design, including drawing it out before you go to the hardware store for parts. To start, decide what size screen you want, and then design the frame to accommodate it. For example, if the screen is going to be 5 by 9 feet (what this project is based on, and a good size for most projectors), then the frame will have to be enough bigger that it can contain the screen cloth stretched taut over the frame, using the zip ties around the outside.

The frame will be a rectangle of 1.5-inch PVC and will require at least three vertical and one horizontal cross-braces. Here's a suggested layout without dimensions (because it's scalable!):

90° ELBOW	PVC HORIZ.	"T" FIT.	PVC HORIZ.	"T" FIT.	PVC HORIZ.	"T" FIT.	PVC HORIZ.	90° ELBOW
PVC VERT.		PVC VERT.		PVC VERT.		PVC VERT.		PVC VERT.
"T" FIT.	PVC HORIZ.	CROSS FIT.	PVC HORIZ.	CROSS FIT.	PVC HORIZ.	CROSS FIT.	PVC HORIZ.	"T" FIT.
PVC VERT.		PVC VERT.		PVC VERT.		PVC VERT.		PVC VERT.
"T" FIT.	PVC HORIZ.	"T" FIT.	PVC HORIZ.	CROSS FIT.	PVC HORIZ.	"T" FIT.	PVC HORIZ.	"T" FIT.
PVC VERT.				PVC VERT.				PVC VERT.

BUILDING THE SCREEN

The blackout cloth should be doubled over and sewn along the edges for reinforcement. Set grommets around the perimeter every foot or so to connect the fabric to the frame. Lay the cloth over the frame and attach it with zip ties. These allow for fine adjustments to get the fabric as centered as possible.

BUILDING THE BASE

To support the frame, sink 3-foot sections of 2-inch PVC in concrete in three 5-gallon buckets. After the concrete dries, slide the screen frame's three 1.5-inch PVC legs deep into the 2-inch pipes for support. A couple of eye bolts at the top sides of the frame allow the screen to be secured with rope and stakes like a tent, which prevents it from moving forward and backward in the event of a breeze. And with that, the screen is done!

To be on the safe side, test the setup to make sure everything worked together. Then you are ready to invite the neighborhood over. My GeekDad suggestion is to kick off the summer by sipping lemonade, eating popcorn, and watching Indiana Jones try to discover the Ark of the Covenant. Happy viewing!

The "Magic" Swing

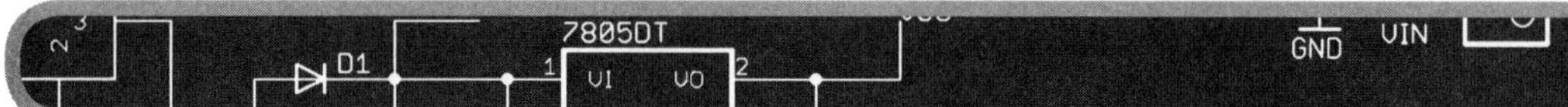

Being a GeekDad means understanding science and, when possible, incorporating that understanding into everyday teachable moments. Whether it's chemistry in the kitchen, dynamics on the pool table, or biology in the garden, improving our kids' knowledge of the world around them is vital.

And an important part of making those teachable moments palatable for them is making science fun. Even when it makes a task a little more complicated, sometimes it's better to add some cool science—to include a "wow" factor in a mundane project—so that it becomes something more.

This project was inspired by that principle and by the geek-favorite show *Mythbusters*.

PROJECT	THE "MAGIC" SWING
CONCEPT	Add a little cool science to an indoor or outdoor swing by including phone books with their pages interlaced as part of the rope line supporting the swing.
COST	$$
DIFFICULTY	⚙
DURATION	⏱ ⏱
REUSABILITY	◍ ◍ ◍ ◍
TOOLS & MATERIALS	Phone books, rope, plywood, eye bolts and nuts, tape, a drill

Friction is one of the fundamental components of dynamics, and yet it can be very difficult to explain beyond sandpaper and rub burns. This project offers a surprisingly cool demonstration of the true power of friction by using it to support the weight of a person on a swing (either your kid or yourself . . . or both!).

WARNING: This is not a practical project. It would be easier to build a rope swing with just rope. But it wouldn't be as cool. Also, keep this out of the rain. The phone books won't last long if they get wet or suffer otherwise extended exposure to the elements. This swing should be built to be easily taken down and stowed away between uses.

What we're doing is taking the classic idea of a rope swing and adding the friction twist by putting two interlaced phone books into the hanging assemblage. If you haven't seen the episode of *Mythbusters*, the simple science here is that, if you take two phone books (and who really needs their phone books these days?) and interlace the pages as you would shuffle a deck of cards, they can't be pulled apart by any practical means.

How does it work? The simple answer is friction. The friction between individual pieces of paper is negligible. However, when you interlace the pages of two phone books with hundreds of pages in each, the cumulative effect is stronger than glue, and you won't be able to pull them apart with anything less than a couple of tanks (hence the "practical" caveat above).

INTERLACING THE PHONE BOOKS

This is the long, slightly boring part of the project. You want to start with two phone books of approximately the same size (identical if you can get them). Set them in front of you on a table, with the binding sides facing out. Imagine shuffling them like a deck of cards: You'd simply pick up the inward-facing, loose sides of the books, bring them together, and shuffle.

Unfortunately, for this project, it takes a little more work than that.

Take the books and fold everything outward, until all you have are the back covers facing inward. Bring the two books toward each other until one cover lays over the other by at least half its width. Then start leafing pages over, first one book, then the other, alternating as you go. Do this until you have interlaced all the pages.

BUILDING THE SWING

Next we're going to incorporate the phone books into the swing. We will build reinforced brackets so we can tie the rope for our rope swing to the spines of the phone books.

STEP 1: Cut 4 strips of plywood (½ inch or thicker) 3 inches wide and as tall as each book to sandwich the un-interlaced ends of each of the phone books.

STEP 2: Take one side of the interlaced phone books and sandwich the spine with two strips of the plywood. Hold the pieces of wood in place with a clamp.

STEP 3: Now drill three holes through the plywood/phone book sandwich, spaced evenly across. The holes will be two different sizes. The center hole should be about the diameter of the rope you're going to use (probably 5/8 inch to 1 inch). The other two should be sized to fit the bolts you're going to use to hold the brackets together, perhaps 3/8 inch or so. Go ahead and drill, and bolt the brackets together, leaving the center holes open on each end of your assembly.

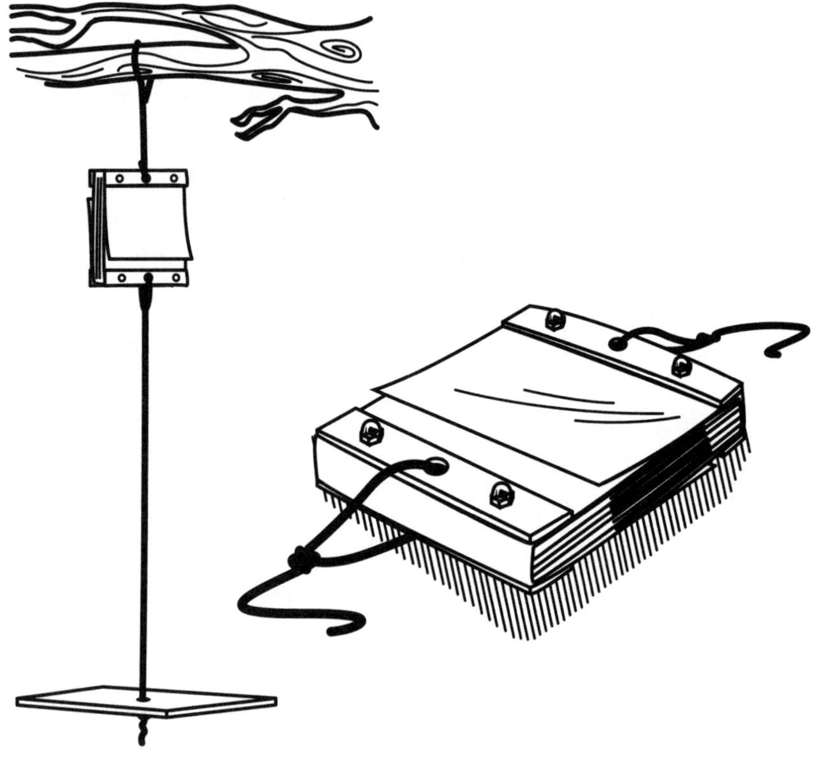

SETTING UP THE SWING

STEP 4: Figure out where you'll hang the swing: off a play set, a sturdy tree branch, or some other overhanging anchor point. Decide how long your rope is going to need to be, roughly. Assume you'll need a knot at the bottom for the seat to rest on, enough length such that the seat will be suspended two to three feet off the ground at the lowest point of the swinging arc, and enough at the top to tie it off at the anchor point. This should be a conservative estimate; you can always trim it later if it's too long.

Your choice of rope is important. You'll need something fairly thick and strong. A heavy climbing rope can be a good choice, as you'll want to be able to knot it relatively easily.

STEP 5: Your first piece of rope will go from the "top" side of the phone book assembly to the anchor point where it hangs. Figure out what you think that distance will be, then triple it to give yourself some length to work with. Feed the rope through the large center hole in one bracket, pull it even so the ends meet when held straight up from the books, and tie a good knot (I'll leave it to your Boy Scout skills or Internet research to determine the right one).

In my version of this build, I hooked the knotted end to a heavy-duty carabineer, and then found a heavy-duty hook I could drill into my anchor point. You may have to adjust depending on your anchor point, available tools, and hardware, but I have faith you can figure it out.

STEP 6: Hang from the anchor what you've got done so far, and try testing it with your weight. Isn't friction amazing?!? Next measure the distance from the bottom of your assembly to where you want the seat to be at the bottom of the swing's arc. You have two choices now: Use either a single length of rope tied off to a carabiner or a metallic ring hooked through the bottom hole of the phone book assembly.

STEP 7: Measure the rope so you have a good foot below the seat point. Build your seat. The simplest is to cut out a square of thick plywood, or sandwich a couple layers of thinner plywood, and drill a hole through the center to pass the rope through. Knot the rope both above and below the seat to keep it in place, making sure the knots are tight and that they're big enough not to pass through the hole.

And with that, you're done! Let your kids try it, even try it yourself, and tell everyone who comes over how the only thing keeping it together is the magic of friction!

(Be sure to bring it in the house when you're done—getting it wet would be bad.)

AWESOME ACCESSORIES

Smart Cuff Links

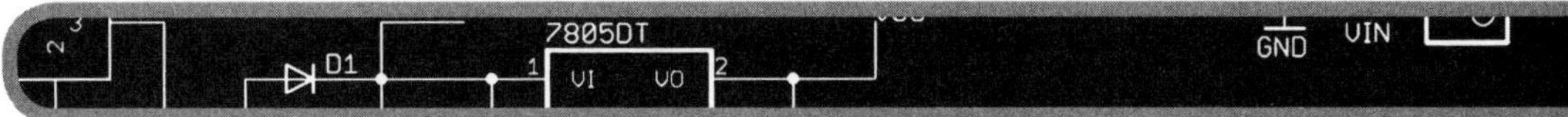

Geek culture tends to be, shall we say, a casual culture. We tend to be more comfortable in khaki shorts and superhero T-shirts than suits—and that's just at work. But there are a few times in a geek's life when fancy dress (and we're not talking cosplay here) may be required. But even if you or your kids have to dress up, that doesn't mean you have to lose your geek cred.

PROJECT	SMART CUFF LINKS
CONCEPT	Make a pair of cuff links out of RJ-45 Ethernet connectors and wire.
COST	$–$$
DIFFICULTY	
DURATION	
REUSABILITY	
TOOLS & MATERIALS	RJ-45 Ethernet connectors, twisted-pair wire, crimping tool

All the parts to this project are available at your local electronics or computer store, though if you're a geek worth your salt, you very likely already have them. Obviously the crimper is the most specialized piece of equipment, but crimpers can be found pretty much everywhere (I've seen them in Home Depot, RadioShack, and Target).

1. For one cuff link, measure out about 3 inches of the twisted-pair wire.

2. Untwist the ends a bit so they'll be easier to get into the connector.

3. Feed one end of the pair into the two available holes on one extreme side of the connector.

4. Feed the other end into the hole on the opposite extreme end of the connector, and crimp per the instructions that come with the tool. This will create about a ¾-inch loop of wire coming out of the back side of the Ethernet connector.

5. Trim a 4-inch piece of twisted-pair wire, fold it in half, and twist together into one thicker length.

6. When you're ready to put the cuff link on, feed the loop from the Ethernet connector through the French cuff in just the way you would the post of a traditional cuff link. Feed the doubled-over 4-inch length through the exposed loop, and twist over so the cuff link stays in place.

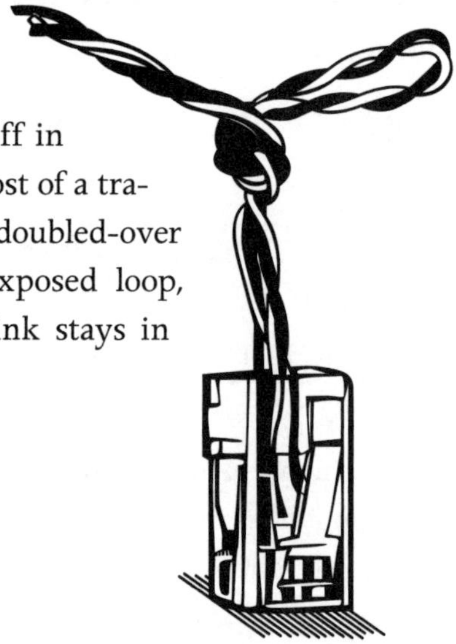

Repeat this with another connector to make a second cuff link to fill out your pair.

An Even Cooler Idea

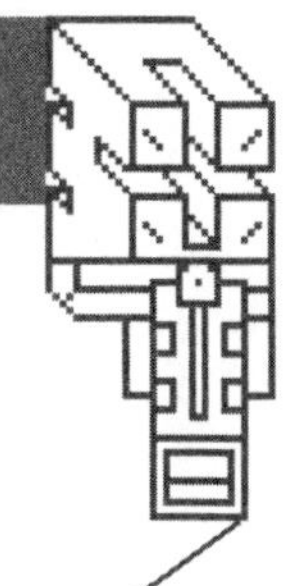

If you are an advanced Ethernet cable installer (meaning you know how to crimp connectors on either end of a length of cable), put an RJ-45 connector at one end of a cable, and the RJ-45 jack at the other. Depending on the length of cable, it could be used as a bracelet, stylish tie, belt, or ID card lanyard.

The end result of this project is a formal accessory for the geek who is truly secure in his technophilia. Whether it's the winter ball or a dance, New Year's Eve, or even a (your?) wedding, this simple bit of whimsy should see you through.

Light-up Duct Tape Wallet

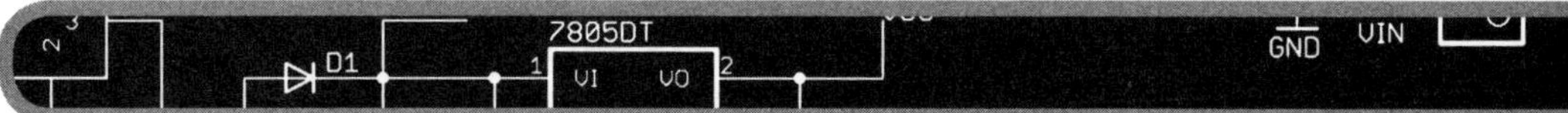

GeekDads tend to do a lot of technical stuff with their kids. Beyond just video games, we love to learn programming, or go geocaching, or hack roadway message boards to flash lines from Tron at passing motorists (okay, maybe that's just me). All very technical geeky activities. And we know geeks can get into crafts as well, from crocheting dice bags (page 157) to cosplay costumes. But what about melding crafts and technical geekery into one project? Oh, that's easy!

PROJECT	LIGHT-UP DUCT TAPE WALLET
CONCEPT	Build a useful, durable wallet out of duct tape, and add a handy electronic feature.
COST	\$–\$\$
DIFFICULTY	
DURATION	–
REUSABILITY	
TOOLS & MATERIALS	Roll of duct tape (you can use more than one color to be creative), box cutter or X-Acto knife, plastic cutting board, ruler, dollar bill, Mini Maglite replacement bulb, AAA battery holder, coated copper wire, mercury switch

Tape as a textile, you ask? Why, of course! Duct tape—another WWII military invention brought to use in everyday life afterward—is fabric-reinforced for strength and has a rubber-based adhesive that resists the effects of moisture, making it a very handy material for a variety of textile projects (there are a slew of Web sites dedicated to various duct tape craft projects). Indeed, the duct tape wallet itself was not invented for the GeekDad book, but the little variation included here was, and with it, we turn the geek factor up from 10 to 11.

There are two halves to this project—the wallet itself and the light module. We'll start with the wallet.

MAKING THE WALLET

To make the wallet, first collect your materials. Of course the primary material is duct tape. For the purpose of this project, we'll stick to just classic silver-backed duct tape, but there are a variety of colors available at your local hardware store, so if you want to get creative in the future, go crazy.

STEP 1: You'll also need a surface to work on. A plastic cutting board is very effective, because while the tape will stick to the board, it can be pulled off fairly easily and still retain most of its adhesive. That will be key for this project. A ruler is not vital since you can work from a dollar bill for your dimensions and eyeball things generally, but a straight-edge is very handy. And if you're a stickler, use a ruler, for I'll give you rough dimensions all the way through. For this part, you'll need that box cutter or X-Acto knife, which will cut through the duct tape like "buttah."

What we're making here is a simple billfold, and you can even use one you have on hand as a guide.

1. Start by laying three overlapping rows of tape on the cutting board to create a rough shape at least 7½ inches wide by 4½ inches tall.

2. Lay your dollar bill (which is 2½ inches tall by 6 inches wide) on top and trim the tape sheet so that it is about ¾ inch wider than the bill on each side, ½ inch clear at the bottom, and 1½ inches on top (see, that makes 7½ by 4½ inches). Try to get nice clean edges.

3. Now carefully peel the whole sheet off the board (use the knife to get under one corner, and then slowly pull the tape away) and lay it adhesive side up.

4. Tape over the sheet with new rows of duct tape, making sure you go at least ½ inch wider all the way around.

5. Trim this cover layer so that it is ½ inch wider and taller, then cut the corners at 45-degree angles (kind of like any piece of paper in *Battlestar Galactica*).

6. Flip the whole piece back over, then fold the overlapping edges over and stick them down so that you end up with a nice sheet of double-backed duct tape—the perfect waterproof wallet material.

STEP 2: Now we'll start on the inner pouch that's going to hold the electronics package (sounds very spy tech, doesn't it?). Our guide for the size of the pouch is the size of the battery holder. I'll talk about the electronics materials later in this project, so you should read the whole thing before starting. However, the battery holder is a simple AAA module like you'd find in many toys or other devices. It's not much bigger than the actual battery, so you can approximate the size of the pouch based on that information.

1. We're going to make a two-sided tape sheet just like the first one, but sized so that, if you rolled the battery holder up in it, the sheet would wrap around it once, with about ½ inch extra (call it 2 inches tall by 3 inches wide).

2. What you're not going to do like the last time is to make a 45-degree corner cut and seal all four sides. Instead, trim the corners square (at 90-degree angles) and seal just the top side by folding the flap over.

3. On the bottom flap, do a little more trimming so it has three serrated-looking tabs. These will become the bottom closure of the pouch.

4. Using the battery holder as your guide, roll the pouch up so that the left side tab seals it up into a tube shape, leaving the right side tab still unadhered.

5. Fold the bottom serrated tabs in on themselves to create the wallet's bottom (there will be a bit of the sticky side facing into the interior of the pouch—this is intentional, to help hold the electronics in place).

STEP 3: You now have a pouch with a sticky tab on the right side, which we'll use to attach the pouch to the inside of the wallet. We'll do that now, by applying it to the first sheet we made. On the left side of the sheet, set the pouch with the tab on the left. The bottom of the pouch should be ½ inch from the bottom of the sheet, and the sticky tab should fold over and behind the left edge of the sheet.

STEP 4: Next we'll make the sheet that finishes our wallet. This sheet is the inside of the billfold, so it should be a little smaller—say 3½ inches. But because it has to allow for the bulk of the pouch inside, it needs to be a bit longer; try 7¾ inches wide. When you've applied the second face, and flipped the sheet back over, trim the top corners at 90-degree angles, and the lower corners at 45-degree angles. Fold the top flap down to seal the top edge.

STEP 5: Now take your first sheet with the pouch and flip it face-down. Line up one lower corner and side edge with your new sheet (make sure the first sheet is oriented so that the open end of the pouch is facing up toward the finished edge of the new sheet). You should be able to fold the side sticky tab of the new sheet over the edge of the old one. Then line up the first sheet's bottom edge with the inside edge of the new sheet so the bottom sticky tab can fold up and over. Finally, line up the other side edges and fold over the last sticky tab so that you have attached the new sheet to the first sheet, with the pouch on the inside.

STEP 6: You now have the basic billfold with a pouch inside. If you want, you can add another inside sheet, perhaps in a different color for a nice accent, and slightly shorter than the inside wall of the billfold, as a pocket for credit cards and ID. Those sorts of additional features I'll leave up to you and your kids' creativity.

BUILDING THE ELECTRONICS

The electronics package is where you can make this really geeky, and where you have the chance to play with the materials you can find. I went with some cool items I dug up at my local Frys Electronics, the handiest of which was the bulb. While we geeks like to dabble in all things LED (as you can well see elsewhere in this book), sometimes the simplest alternative can be the best and easiest to deal with. When I saw these replacement bulbs for Mini Maglites that were specifically designed for use with a single AAA battery, I was happy. What's easier than battery/bulb/switch to build a circuit for something like this?

Admittedly, this could be done a bit geekier, and more streamlined, with LEDs and a CR2032 battery, but I'll leave that up to you to find and figure out.

Where I did go geeky was with the switch. While there are numerous neat little switches available on the aisle at my store, when I saw they had a small selection of mercury switches, I just had to use one!

Assembly of the electronics is a commonsense process, building a circuit from the negative end of the battery holder, through the switch, the bulb, and then to the positive end of the holder.

1. Take the battery holder (sans battery for the time being) and attach one lead of the mercury switch to the negative end (usually the end with the spring). You might want to use a small pair of needle-nose pliers to help manipulate the wiring, and make sure you wrap the wire up in such a way as to keep things from shorting.

2. You can use small strips of duct tape to hold the switch to the side of the battery holder. Make sure the mercury switch is positioned so that, when the wallet is "up"—meaning that the open side for looking in to see the contents is facing toward the sky—the switch is activated (the blob of mercury is on the "bottom" of the bulb where it touches the two contacts to close the circuit).

3. The other lead of the mercury switch is attached to a small length of wire, so the circuit can reach the top/positive end of the battery holder.

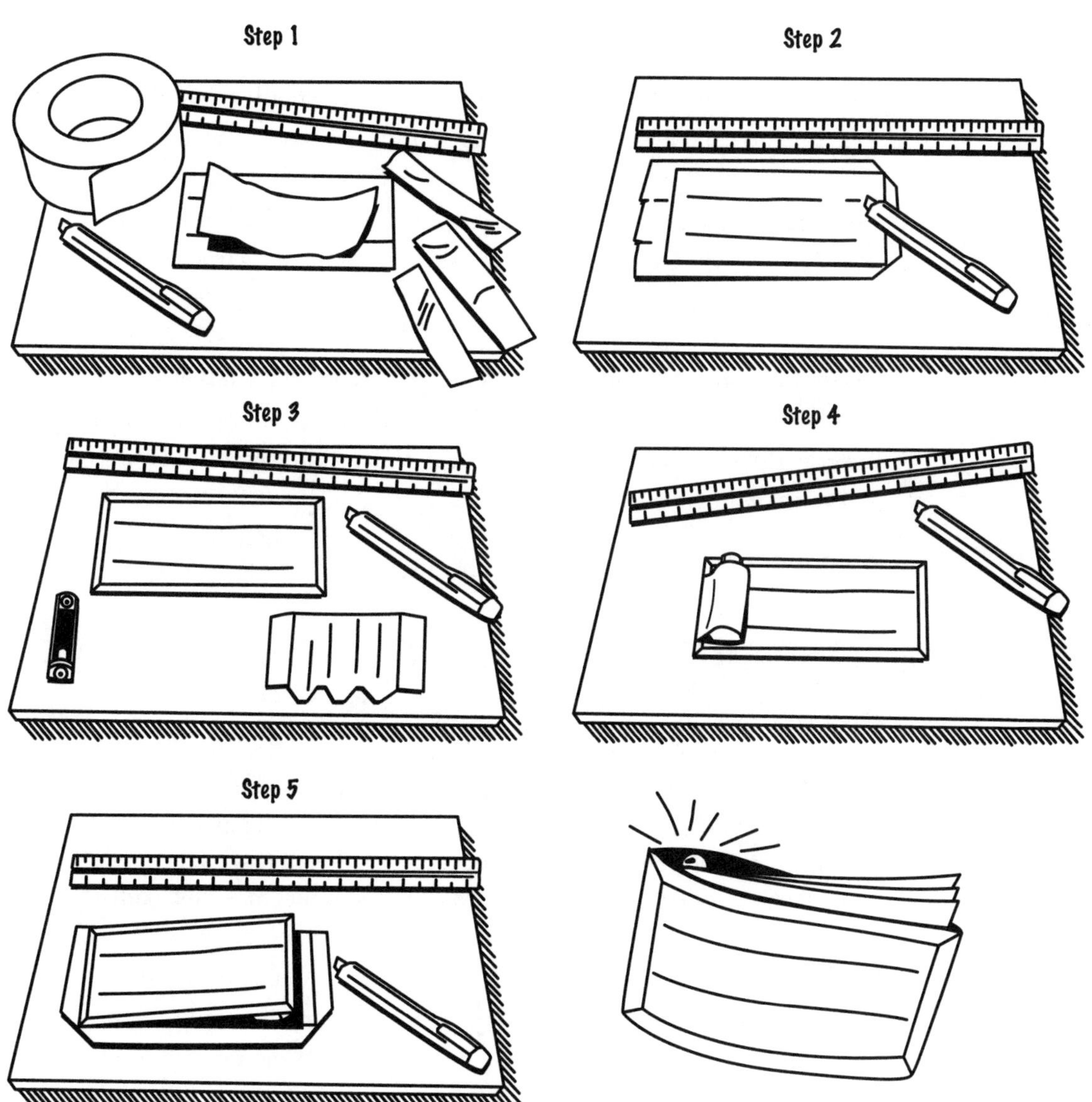
Step 1
Step 2
Step 3
Step 4
Step 5

4. At the top, connect one lead of the lamp to the wire, and the other to the positive lead of the battery holder.

5. A few twists of the wires, and some more small strips of duct tape to hold everything together, and you have a circuit.

6. Put the AAA battery into the holder, and then test it by turning it upside down and right side up to make sure the bulb turns off and on.

Just slip the package into the pouch inside your wallet. The bulb should just stick out of the top of the pouch, but should not clear the edges of the wallet itself. Put a few dollars into the billfold, and marvel at your lighted wallet!

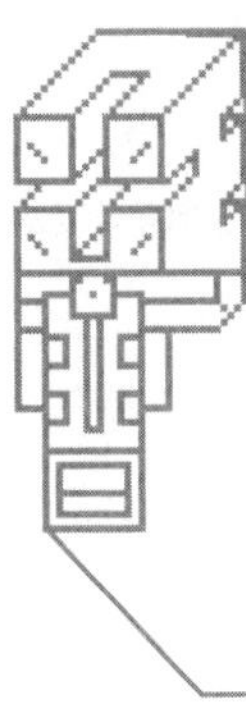

Please Remember!

There are two caveats that go along with this design. First, for safety, this should be a front-pocket wallet, since both the bulb and the switch could be crushed if you sit on them. Second, by using the mercury switch, when the wallet is in your pocket, it needs to be top side down so the light stays off. Thus, keeping coins in the wallet may be a bad idea.

Crocheted Dice Bag of Holding

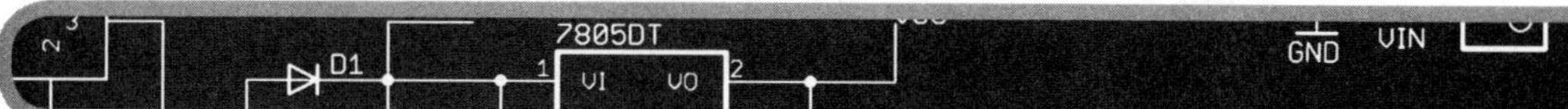

One of the most important accoutrement for any good RPG-playing GeekDad or kid is the dice bag. It is the container for your most trusted tools: d20, d12, d10, d8, d6, and d4. Indeed, you may have more than one, since you may have different sets of dice for different games (the full range for D&D, all d6s for old-school Champions, and so forth). And if you want to treat your tools right, as well as showing off your dedication to the games you play, you'll want to make your own bag.

So it's time for you and your geeklet to get in touch with your crafty side again, with this great project from GeekDad writer and serious RPG player Natania Barron:

PROJECT	CROCHETED DICE BAG OF HOLDING
CONCEPT	Crochet your own RPG dice bag.
COST	\$–\$\$
DIFFICULTY	
DURATION	
REUSABILITY	
TOOLS & MATERIALS	Crochet hooks, yarn

Crochet is definitely a geek-certified activity. It's important to understand these key concepts: Crochet is basically all about knot tying, which is a geeky pursuit from time immemorial. You are tying knots into yarn to build a structure; indeed, you could think of it as a kind of textile engineering! On top of that, crochet patterns aren't really patterns at all: They're programs. You are processing lines of code, and what's geekier than that? Besides, every gamer young (your kid) and old (you) has to have a cool and original dice bag for the gaming table. Why not make your own?

GeekMom Natania was once a big knitter, but when she had her son, it became abundantly clear that knitting just wasn't an option for her. Her son's deft little fingers easily destroyed hours of work in seconds, and left her frustrated (and out a few Christmas presents). She knew about crochet, but thought it was, you know, for gray-haired ladies crouched around coffee tables at nursing homes. Thankfully, with Web sites like www.ravelry.com (a good place to start for basic crochet tips and tricks, and to grow into the community), as well as a recent resurgence in the art of crochet, she learned better.

The huge benefit of crochet is that the whole work rests on one loop. So, unless your child is being really devious, it's unlikely that they can undo the work as quickly as with knitting. And the added benefit is that crochet is super malleable. It's almost like working with clay. Once you learn the basics, you can make all sorts of projects that, as with knitting, would be a lot more challenging—hats, socks, toys—truly endless possibilities.

This dice bag came out of Natania's desire to add a little style to the RPG gaming table, where the dice container is just as important an accoutrement as the dice themselves. Using the bits and pieces of yarn left in her knitting stash, Natania worked out a dice bag that would be roomy enough for a fistful of dice, yet still be sturdy. After a few prototypes, the best design appears to be made out of 100 percent wool, felted, with a totally flat bottom. The flat bottom works well because it rests on the gaming table perfectly, so you don't have to spill dice all over the place and can just peek in and grab what you need. That said, the pattern allows for plenty of alteration. You can add colors, vary the stitches, or get really saucy and bead the bag.

Here's the "program" for building your dice bag (Asterisks surround instructions that get repeated):

- To Start: Chain 3. Slip stitch into 1st chain.
- Row 1: Double crochet into center loop eight times. Sl st into 1st dc.
- Rows 2-3: Ch 2. *2 dc into each dc.* Repeat around. Sl st into top of 1st dc.
- Row 4: Ch 2. *Dc into 1st dc. 2 dc in next.* Repeat around. Sl st into 1st dc.
- Row 5: Continue increasing as in row 4 around the circle until the base is the desired width. (I typically like the size at about the fifth row.)
- Row 6 and all following rows until desired height is reached: Ch 2. *Dc in each dc.* You can either crochet in the round, or work with rows. I prefer rows because the top is smooth.
- Final Row: Ch 2; *Dc into first dc. 2 dc into 2nd.* Repeat all the way around. This adds a little lip, or flare, at the top, good for tying.

If you are felting (a process that teases out the fibers of the wool yarn in your crochet to give it a fuller, fabric-y look), don't worry about weaving in ends; you can snip them off later. If you are not felting, weave in the ends so there's no loose tails.

To felt: Toss the bag into your washer. If you aren't worried about specific sizing, just leave it in there, letting it swish around. Make sure the setting is hot and on high agitation. When the cycle's done, your stitches will be dissolved. To shape, pull the still wet—

don't dry this in the dryer—bag over a mug, cup, or an ice bucket from a drink set (really? An ice bucket?).

Make sure if you are using multiple colors that the skeins are the same brand of yarn, and steer clear of white since it doesn't felt well. Unless you want to vary felting textures; by all means, work it up however you like.

There are a few options for the tie. You can make a double-tie system, using two laces threaded at either side to pull the top taut. This allows the bag to close tightly without actually tying the laces. But sometimes a simple lace does the job perfectly; just poke right through the felted material to wind the laces. To make a lace, you can simply chain wool or, for a smoother option, nylon.

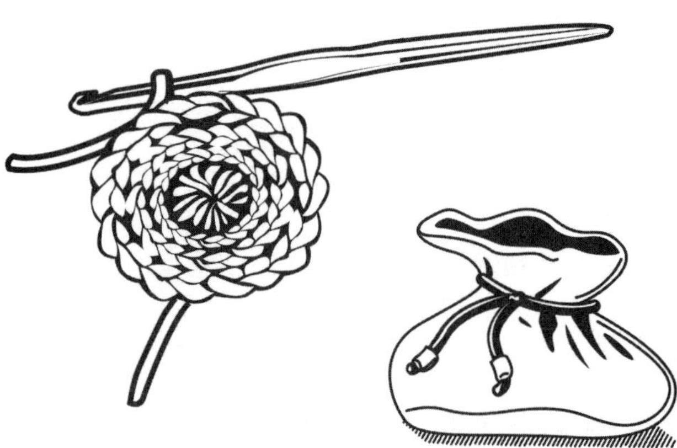

If you're feeling funky, you can always embellish with beads. Or, you know, Dremmel through some d8s and use those. . . .

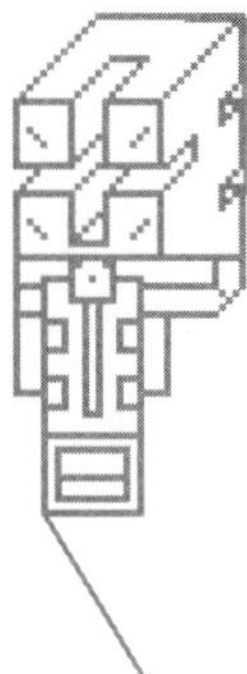

An Even Cooler Idea!

Once you've mastered crochet and it's kissing cousin, knitting, there's a whole world of geeky craft projects waiting for you. Some other items you can make, as suggested by J. Lynne on the D3blog (http://jlynne.exit-23.net/2007/12/05/thirteen-geeky-knitting-projects/), include Jayne Cobb's hat from firefly, an R2D2 Cap, a Princess Leia Wig-Hat, Tom Baker's Doctor Who Scarf, a Hogwarts Gryffindor Scarf, or anything else you and your geeklet's imagination can come up with.

GEEKY KIDS
GO GREEN

The Science of Composting

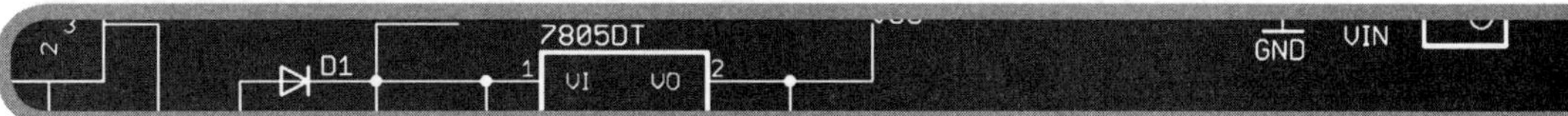

As I write this book, my home in the San Jose/San Francisco/Oakland Bay Area has been experiencing the coolest summer in my memory. One might say "strange things are afoot at the Circle K," for the logical set where "Circle K" = the environment.

Geeks are into science, and the overwhelming scientific evidence is telling us that the effects of our environment-unfriendly activities, starting with the Industrial Revolution and continuing through today's manufacturing, motoring, and mining have accumulated to the point where they are having macro-environmental results. As geeky parents, it behooves us to educate our kids about this evidence and the things that need to be done to halt our downward spiral. One very useful way of doing that is by being environmentally conscious at home.

But being environmentally conscious on a personal level can mean many things. For example, our boys don't receive a regular allowance. Rather, they get to split the redemption money from every can and bottle we accumulate and take to our local recycling center.

Saving energy is also important, though we're still working hard to get them to turn off their game consoles and lights when they leave the room.

But one more activity can encourage a more environmentally friendly (not to mention healthier) lifestyle and provide some important education—and that's growing food. While farming may not seem terribly geeky, I assure you that, when you add science into the mix, anyone can be a geeky gardener. Which is why I've included, in this chapter and the next, two easy and slightly geeky home-gardening-related projects you and your geeky children can build together.

PROJECT	THE SCIENCE OF COMPOSTING
CONCEPT	Build a compost bin out of simple materials.
COST	$$
DIFFICULTY	⚙
DURATION	☼–☼☼
REUSABILITY	◍◍◍◍
TOOLS & MATERIALS	Plastic tub, power drill, rocks, shredded newspaper, compost starter

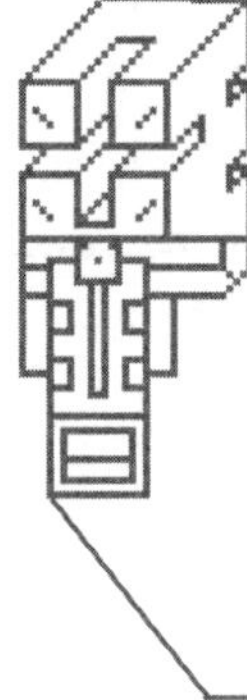

A Note on Materials

You can pick up a plastic storage tub from your local big-box store. It doesn't need to be anything fancy—just one of those 20-gallon bins with a lid if you're starting small, on up to a full-sized garbage can if you are a bit more ambitious. The compost starter is slightly more specialized (though not as rare as it used to be). You should be able to find it at your local garden supply store, or online.

STEP 1: With your close supervision, have your child drill a series of holes in the top (lid) and bottom of the bin. There's no hard and fast rule on the size, quantity, or spacing, but assume something like 8 to 12 holes with a $\frac{1}{4}$-inch bit, evenly spaced, both top and bottom. This should allow an adequate level of aeration and seepage.

STEP 2: Fill the bin about 6 inches up from the bottom with a mix of shredded newspaper and small rocks (the rocks aren't essential, but they'll help things along). Then add enough compost starter to get to at least the half-full mark.

STEP 3: Before we get the composting actually going, you need to ask yourself a question: Where is the best location for your composting bin? Consider that it is utilitarian in appearance, it may seep a little, and it will have a tendency to smell a little at times. Plus, you'll want easy access to it for dumping new material into it on a regular basis, and you'll need to be able to water it from time to time. You may even want to think toward the future and put it somewhere that has room for expansion in case you really geek out over this composting business.

About Compost Starter

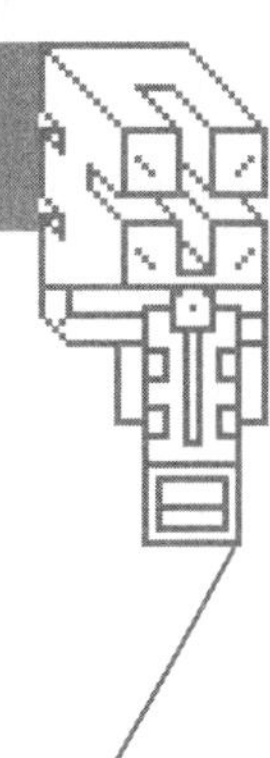

You don't have to use specific compost starter for this project. You can simply start with good soil instead, but the composting cycle will take a lot longer to really get rolling. You also don't have to buy it from a store. Check around with your neighbors to see if anyone else is composting, and if they are, ask if you can have some of theirs as a starter for your bin. Most folks cool enough to be composting will probably be cool enough to share.

STEP 4: When you have your bin properly situated, it's time to start filling it up. There are a number of very useful resources for composting information online (see the links in Appendix A), but here are some guidelines for what you can and can't compost:

THINGS YOU CAN COMPOST

Apple cores, Aquarium plants, Artichoke leaves, Banana peels, Bird cage cleanings, Bone meal, Bread crusts, Brewing wastes, Brown paper bags, Burlap coffee bags, Burned toast, Cardboard cereal boxes (shredded), Chocolate cookies, Citrus rinds, Coconut hull fiber, Coffee grounds, Cooked rice, Crab, shrimp, and lobster shells, Date pits, Dead bugs, Dried-up and faded herbs, Dust bunnies, Egg shells, Elmer's glue, Expired floral arrangements (sans ribbons), Feathers, Fish bones, Fish meal, Fish scraps, Flower petals, Freezer-burned fish, Freezer-burned fruit, Freezer-burned vegetables, Fruit salad, Grapefruit rinds, Grass clippings, Greeting card envelopes, Grocery receipts, Guinea pig cage cleanings, Houseplant trimmings, Ivory soap scraps, Jell-O (gelatin), Kleenex tissues, Leather wallets, Leather watchbands, Leaves, Lint from clothes dryer, Liquid from canned fruits and vegetables, Macaroni and cheese, Matches (paper or wood), Melted ice cream, Moldy cheese, Most food waste, Nut shells, Old beer, Old leather gardening gloves, Old or outdated seeds, Old pasta, Old spices, Old yogurt, Olive pits, Onion skins, Paper napkins, Paper towels, Peanut shells, Pencil shavings, Pet hair, Pickles, Pie crust, Pine needles, Unpopped popcorn, Post-it notes, Potato peelings, Produce trimmings, Pumpkin seeds, Q-tips (cardboard, not plastic), Sawdust (too much may slow the process), Shredded cardboard, Shredded newspapers (avoid the glossy circulars), Spoiled fruits and vegetables, Stale bread and baked goods, Stale breakfast cereal, Stale potato chips, Tea bags and grounds, Tofu, Weeds, Wool clothing.

THINGS YOU CAN'T COMPOST

Animal wastes, Chemically treated wood products, Diseased plants, Fatty foods, Meats, Meat bones, Nonorganic materials, Pernicious weeds, Plastics (unless explicitly labeled as biodegradable).

Wow. One list surely is a lot bigger than the other, isn't it?

Home Hydroponics

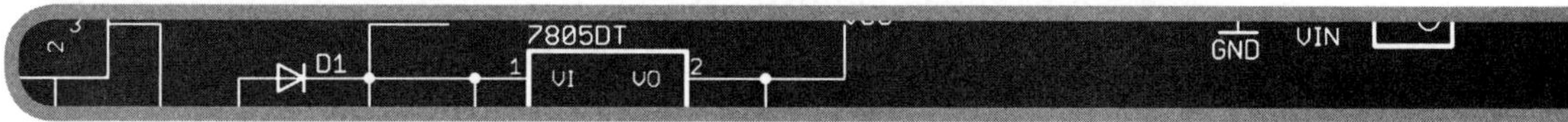

Sometimes you just don't have the space or appropriate location for a composting bin, let alone the dedicated garden to feed with the results. But you can still enjoy the exciting science of gardening. This second geeky green thumb project will let you bring the gardening indoors.

PROJECT	HOME HYDROPONICS
CONCEPT	Build an indoor hydroponic food garden using repurposed materials.
COST	$-$$
DIFFICULTY	⚙⚙
DURATION	⏱–⏱⏱
REUSABILITY	🌐🌐🌐🌐
TOOLS & MATERIALS	2-liter soda bottle, scissors or cutting knife, drill, hydroponic nutrient, food plants, pH test kit or litmus paper, lemon juice, baking soda, aquarium pump, air stone (a little cap of porous rock that helps make lots of really tiny bubbles rather than a few larger bubbles—better for oxygenating the water. You should be able to find air stones at any pet store that sells fish tank supplies), grow lights

Hydroponics is the science of growing plants (often food) without soil or direct sunlight. It's great for homes without yards, or when you want to grow plants under the most protected conditions possible. It's also less messy than digging around in the dirt, and depending upon how you set things up, it just looks cool.

Even better is the fact that it's very easy and inexpensive to build your own hydroponic garden, with a mix of store-bought and recycled materials. And the first time your family sits down to a meal with food you and your kids grew in the garage, everyone will smile a little geeky smile.

STEP 1: A typical 2-liter soda bottle is 12 inches tall (bet you didn't know that). You'll want to cut yours horizontally 5 inches up from the bottom (or, you know, 7 inches down from the top), so you'll have a top piece that's slightly longer than the bottom. If you turn the top piece over and insert it into the bottom, the top of the cap should be able to come to rest on the raised center of the base.

STEP 2: Take the cap off the top piece and, on a workbench or other safe area, drill a series of holes in the top of the cap. If you use a ⅛-inch drill bit, you should be able to get at least five evenly spaced holes in the cap. When you have your holes, put the cap back on the top piece of the bottle.

STEP 3: Feed the hose from your pump, with the air stone on the end of it, into the bottom half of the bottle (which we will henceforth call the "reservoir"), and slip the top half (now and evermore the "container"), back in so the cap touches the raised center of the base. The pump hose should feed up the side between the inner wall created by the container and the outer wall created by the reservoir.

A Note about Pump Choice

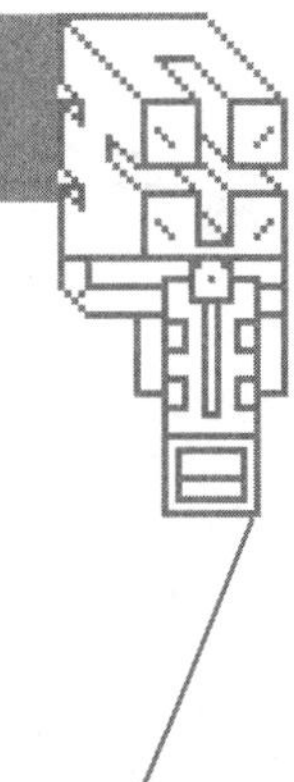

If you are doing just one setup, you can probably get away with the lowest-capacity (and therefore cheapest) pump you can find. However, if you want to build a grand hydroponic garden facility, you can buy a somewhat higher-capacity pump, and then purchase gang valves—essentially controlled splitters—which will let you feed the air from one pump to many hoses. Indeed, you could probably feed gang valves into gang valves to feed many setups, since the aeration requirements of just one reservoir/container setup are quite low.

STEP 4: If you plan to use tap water for your setup, you'll need to test the pH and adjust it to as close to 7 as possible with the lemon juice (if the pH is too high) or baking soda (if it's too low). You'll have to experiment with concentrations if you do this. Of course your alternative is to use distilled water, costing a bit more but saving some time and effort.

STEP 5: Mix your water and nutrient solution per the instructions on the bottle, and pour enough into the setup so that the reservoir fills about halfway.

STEP 6: Now you need to add your growing material—basically the non-dirt "soil" that the roots of your plants will grow in. There are a whole bunch of choices here, from cheap to pricey. At the low end, some kind of gravel, glass beads, or even (yes, it's been done) LEGO bricks will work. The roots of your plants just want to grow through something, and any block of porous material will work. If you want to kick it up a notch, the hydroponics supply sites have perlite "rocks," a material that holds moisture and encourages root growth, or even pellets made of coconut hair. It's up to your budget and geeky obsession how far you want to take this.

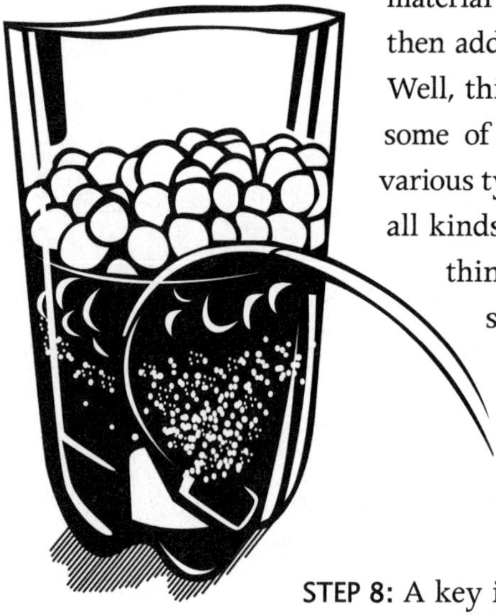

STEP 7: Fill the container up to a point where the surface of the material is also the surface of the water/nutrient mixture, then add your plant. What plants should you be using? Well, this project is intended to be food-producing, and some of the best food plants for hydroponics include various types of lettuce and other green, leafy vegetables, all kinds of herbs, and tomatoes. Of course, with something like tomatoes, you may need to work out a staking system so the tomato vines can grow upward and support the weight of the fruit when it matures.

STEP 8: A key ingredient to indoor plant growth is lighting. If you have an available window that lets in good sun for part of the day, and that you don't mind blocking with growing vegetables, you can put your setup there. But if you're doing the indoor thing all the way (like in the garage), you'll need dedicated lighting. Fluorescents work, especially if you pick up bulbs that generate simulated "full daylight." It's nice that there are also compact fluorescents available to use as well. They aren't cheap, but if you're going to leave these on all day and all night, the savings in electricity and replacement costs will make it worthwhile in the long run.

There are more sophisticated hydroponics systems available—indeed, this is another one of those hobbies into which you can sink a lot of money. However, the basics are pretty straightforward, can be achieved quickly and inexpensively, and the rewards keep on giving. With just a little quality time doing this project with your child, you can have fresh organically grown salad fixings available every day!

BUILD/LEARN/ GEEK

Build a Binary Calendar

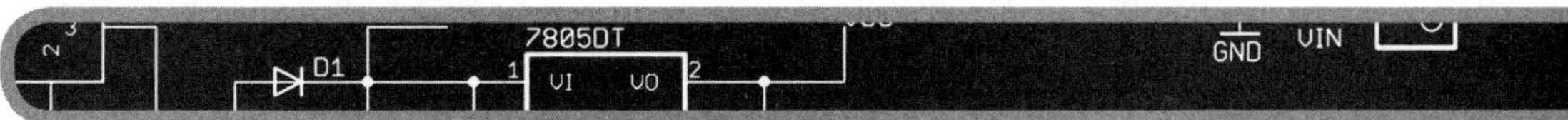

One mathematical concept that's key for any kid who is interested in digging into computers is the binary numeral system. It's at the heart of how digital anything is done. A bit represents the two possible states of a switch: 0 or 1, off or on. Eight bits make up a byte (usually; there are plenty of exceptions, but they're more technical than I want to go into here), which computers use to deal with language. Eight switches that can be either on or off can represent 256 different characters (two states, on or off, times two states, times two states, and so on; or $2^8 = 256$). Kind of like an old-fashioned representational code, 256 numbers are mapped to all the letters (uppercase and lowercase), digits, punctuation marks, and so on, so we can save and manipulate our language on a machine.

All that's pretty basic Computer 101 stuff, but the idea of binary is still a small speed bump to get over when starting out. Which is why I've included this next project, as a means to get your child thinking like a code monkey every day.

PROJECT	BUILD A BINARY CALENDAR
CONCEPT	Build a simple manual calendar that uses binary to tell the date.
COST	$
DIFFICULTY	⚙
DURATION	☼–☼☼
REUSABILITY	⊕⊕⊕⊕
TOOLS & MATERIALS	Basic LEGO bricks

We are going to make a simple LEGO structure that will let you and your child track the day, month, and date manually in binary. What does *manually* mean? Sort of the way the set of calendar blocks at the bank counter works—someone sets the date manually with them each morning. No, it's not very high-tech. That's not the point.

First, how are we going to represent the day, month, and date in binary, you ask? Well, it'll be a representational code, as I mentioned above. A given binary sequence, translated to its base 10 equivalent, will represent the calendar item. For example, the day of the week. There are seven days of the week: Monday, Tuesday, and so forth. If we represent each one with a number, 1 through 7, we can now translate that into binary. Since 7 is less than 8 (duh!), we can represent the day of the week in a 3-digit binary number, for which there are 8 (2^3) possible variations—1 in binary (001) represents Sunday, 2 in binary (010) is Monday, and so forth.

For the numbers 1 through 7, which will represent the days of the week Sunday through Saturday, here is the day-to-decimal-to-binary relation:

DAY	SUN	MON	TUE	WED	THU	FRI	SAT
DECIMAL	1	2	3	4	5	6	7
BINARY	001	010	011	100	101	110	111

If we're going to do the same with months, we have to expand a little. There are 12 months in the year, so we're going to need a bigger binary number set to count up to 12. This time we'll use a 4-digit binary number set (2^4 gives us 16 possible combinations, of which we'll only use 12). See if you can see the pattern in the 3-digit set.

	JAN	FEB	MAR	APR	MAY	JUN	JUL	AUG	SEP	OCT	NOV	DEC
DECIMAL	1	2	3	4	5	6	7	8	9	10	11	12
BINARY	0001	0010	0011	0100	0101	0110	0111	1000	1001	1010	1011	1100

And last, we have to do the same with the numerical day of the month. Since there are up to 31 days in any given month, we'll need a 5-digit binary number this time (2^5 gives us 32 possible combinations, which is just enough to cover our needs).

DATE	1st	2nd	3rd	4th	5th	6th	7th	8th
DECIMAL	1	2	3	4	5	6	7	8
BINARY	00001	00010	00011	00100	00101	00110	00111	01000

DATE	9th	10th	11th	12th	13th	14th	15th	16th
DECIMAL	9	10	11	12	13	14	15	16
BINARY	00001	01010	01011	01100	01101	01110	01111	10000

DATE	17th	18th	19th	20th	21st	22nd	23rd	24th
DECIMAL	17	18	19	20	21	22	23	24
BINARY	10001	10010	10011	10100	10101	10110	10111	11000

DATE	25th	26th	27th	28th	29th	30th	31st
DECIMAL	25	26	27	28	29	30	31
BINARY	11001	11010	11011	11100	11101	11110	11111

Now that we've got the code, what are we going to do with it? Well, what we want to build is something that lets us easily differentiate between two choices: 0/1, off/on, black/white—any pairing that makes it obvious that one of two possibilities are being displayed to demonstrate a binary state. There are a number of variations on this project online, the simplest of which uses pennies faceup, or facedown, to represent 0 or 1. But we like LEGO bricks. With LEGO bricks, there are some simple options available: either colors or face. And to make things really visual, we're going to use both!

Here I'll acknowledge we're really stretching you and your kids' quick translation abilities with this project. The end product will require quick translation from a color/surface combination to the binary digit it represents; then translation of a group of such digits to their decimal version, and from there to their calendar meaning. Pretty tricky! But it's doable, and a great training device for a sharp mind.

This will be a pretty simple build as LEGO projects go. It makes use of pretty standard pieces, and if you don't have them all on hand, you can improvise. Or even better, there is a downloadable file of the LEGO Digital Designer file for this project on www.geekdadbook.com, which you can take to the online store at www.lego.com to order the pieces you'll need, or print out and take with you to your local LEGO store to buy the loose pieces.

My version of the calendar is made on a 12-by-24 base plate. It's somewhat of a nonstandard piece, and if you can't find one easily, you can user smaller plates, overlapped and built up to get the same size and shape.

All we're doing is creating a double layer of black bricks that cover the entire base plate except for the places where our 2-by-2 bricks will fit to be used as the binary counters. In each of those slots, there will be a 2-by-2 base plate with a 2-by-2 smooth-topped plate at the bottom. This will allow us to slip the binary counters in and out easily.

Make sure you put the smooth plates on the 2-by-2 plates before you start building the second level, or you're going to make things harder for yourself later on.

Once you have the base set up, you'll make your binary counters—the easiest part, really. Just pick sets of contrasting 2-by-2 bricks, and make two-layer stacks. You'll need three pairs of one color combination, four of another, and five of a third combo. You're probably way ahead of me here, but these are going to be our 0/1 choices for our three different binary numbers that will represent the day, month, and date as we worked through above.

So now you have to decide, for a given set of color-combined blocks, which color is going to be 0 and which will be 1. Whichever you decide to be 0, make that the bottom brick, so that when the pair is upside down, you actually see the circle shape that looks just like a zero, making it easy to remember.

With all that figured out, assemble your counters and put them in their appropriate slots on the appropriate rows, all upside down (or 0) to start with. Then figure out what day it is, and use the charts above to figure out what the calendar should read. For example, the day and date of the writing of this chapter is Sunday, May 17th. Sunday is the 1st day of the week, so the decimal number is 1, and the binary number is 01. May is the 5th month, so the decimal number is 5, and the binary number is 0101. The date is the 17th, so the decimal number is 17, and the binary number is 10001. And the calendar readings should look like this:

0 0 1
0 1 0 1
1 0 0 0 1

With a few weeks' worth of practice, reading and adjusting the calendar should become second nature, and your geeks-in-training will have gained one more cool math skill that will hold them in good stead in the future.

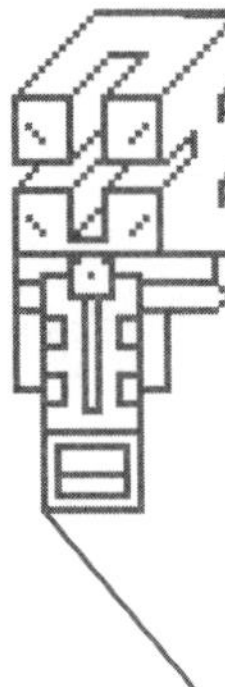

An Even Cooler Idea!

Of course, if you really want to make this interesting, build a Mindstorms NXT robot that uses the color sensor to identify the counters and automatically updates the calendar each morning. Since the stock kit comes with instructions on how to use the color sensor to sort colored balls, it shouldn't take too much geeky engineering with your kid!

Portable Electronic Flash Cards

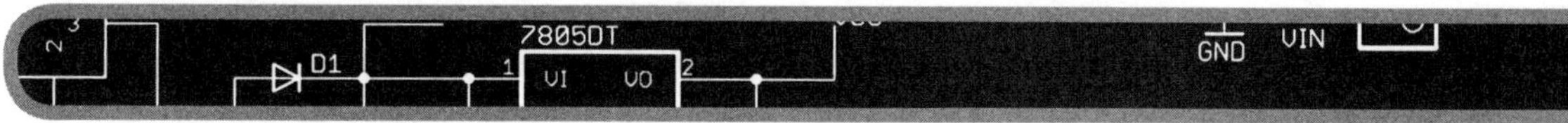

Back in the day, there was nothing more intrinsic to the education process—especially the part of the education process that favored rote memorization over a holistic understanding of ideas and problem solving—than the flash card. They were, and continue to be, a simple idea: 3-by-5-inch rectangles of heavy paper stock, often with lines on one side, and available in a rainbow of colors. You write the question/problem/buzzword on the blank side, and the detailed data or answer on the lined side. Then you take the cards and run through them, over and over, testing and confirming or correcting until the whole stack of learning has been transferred (hopefully) into your long-term memory—or at least into your short-term memory long enough to pass that upcoming test.

These days, save for the hipsters and their Luddite PDAs, paper is passé, and toting around sets of cards for the sake of memorization seems so . . . quaint. But the function of first creating the cards (tactile learning through writing the data), and then repeated self-testing (visual and verbal learning), is still a valid technique. And it's not as if our kids aren't already used to carrying around portable information transference systems everywhere they go—Nintendo DSes or Sony PSPs, various Web-browsing smartphones, and so on.

So what if they could carry around their flash card decks on their portable entertainment devices? Well, why not?

PROJECT	PORTABLE ELECTRONIC FLASH CARDS
CONCEPT	Build flash card "decks" using a JavaScript file, and install it on your kids' portable devices for quick use anywhere.
COST	$
DIFFICULTY	
DURATION	
REUSABILITY	
TOOLS & MATERIALS	A computer, a portable electronic device (Sony PSP, iPhone, iPod Touch, other smartphone with browser)

Many portable devices these days have the means to browse the Internet, meaning they have some kind of Web browser installed, whether it's Safari on the iPhone, the mobile version of Internet Explorer on Windows Mobile devices, or the browser on the Sony PSP. What this means is, they can view HTML files and can usually also read the JavaScript language. The goal of this project is to give you a JavaScript file you and your child can edit (even if you're not a true "coder") to create your own decks of flash cards to carry and access anywhere on a portable device.

But let's take a step back to the beginning. This project started with my friend Bill Moore, a buddy of mine since high school and a groomsman at my wedding. Bill is a good geeky dad in his own right. He would faithfully help his son prepare decks of flash cards to learn school material, but he was troubled by the inefficiency and downright environmentally harmful nature of the task. The cards were usually of value only for a short time—once the subject was quizzed and done with, the cards were not useful any longer and they became a terrible waste of money, effort, and trees. He and his son tried writing them in pencil so they could erase the text and then reuse the cards, but that became very annoying very quickly. They didn't want to abandon the usefulness of the flash-card method of learning, but they needed to figure out a better way of doing it.

Then one day, Bill had a brilliant idea. Considering that his son carried his Sony PSP just about everywhere he went, Bill realized doing something electronic could be the solution. Since it was what he was familiar with, he started with a Perl script that would generate a file viewable on the PSP's Internet browser, and then worked out how to save and access that file in the root directory of the PSP's internal storage. The proof-of-concept worked, but it was a little ungainly for repeated use, so he did some research and figured out JavaScript could be the answer he was looking for. With a bit of work, and about 30KB of coding, he had a functional script with which to build "decks" of electronic flash cards.

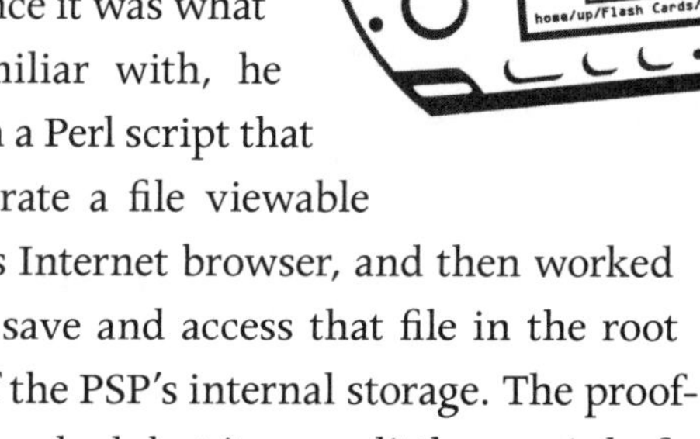

The full script is available online for free, as an open-source project, at www.geekdadbook.com.

Most of the code builds the look and functionality of the tool (features which, if you are a scripting guru, you are more than welcome to improve and share), but the important part to understand is this stretch:

```
function initCards() {
newSubject("Science");
newChapter("Ch 3—Matter");
addCard("matter", "anything that takes up space, has mass, and
    has properties that you can observe and describe");
```

```
addCard("mass", "the amount of matter making up an object");
addCard("volume", "how much space an object takes up");
addCard("weight", "the measure of the pull of gravity between an
    object and Earth");
addCard("density", "the amount of matter in a given space");
addCard("element", "a substance made up of only one type of
    matter");
addCard("atom", "the smallest particle of an element");
addCard("mixture", "two or more types of matter that are mixed
    together; each keeping its own chemical properties");
addCard("filter", "a tool used to separate things by size");
addCard("evaporation", "the change of a liquid to gas");
addCard("compound", "a substance made when two or more
    elements are joined, each losing its own properties");
addCard("buoyancy", "the upward force of water or air on an
    object");
addCard("solid", "a form of matter that has a definite amount of
    space");
addCard("liquid", "a form of matter that takes up a definite
    amount of space and has no definite shape");
addCard("gas", "a form of matter that does not take up a definite
    amount of space and has no definite shape");
addCard("metric system", "a system of measurement based on
    units of 10");
addCard("length", "the number of units that fit along one edge of
    something");
addCard("area", "the number of unit squares that fit inside a
    surface");
endChapter("Ch 3—Matter");
endSubject("Science");
}
```

This is where the actual data is stored. Everything between "function initCards() {" and the last "}" is the contents of the deck. The structure is pretty simple: "newSubject" creates a new subject listing for a stack of cards on a particular topic (in this case, Science—with the quotes); "newChapter" starts a sub-subject (or, um, chapter), and then "addCard" starts a new card. In between the parentheses goes the front side text, in quotes, a comma, and the back side detailed info, in quotes. Like this:

```
newSubject("subject name");
    newChapter("chapter title");
    addCard("front side text", "back side text");
    endChapter("chapter title");
    endSubject("subject name");
```

What's very cool is that you can do as many subjects, as many chapters, and as many cards as you want, simply by repeating the basic script over again on new lines and changing the titles. You could, if you and your geeklet wanted to, have a stack for every school subject (say Science, Math, English, Foreign Language, Social Studies). In each of those, create a chapter for every chapter in his textbook for the given subject, and then as the school year progresses, add the cards for a given chapter as it was learned in class, so that by the end of the year, the deck would represent all the key facts your child learned, and be the ultimate study guide for finals.

But more important, once you and your kids have learned how to build the electronic cards, work with them to choose what goes on a given deck, but have them do the coding. Just like making handwritten cards in the "old days" provided half the educational benefit (tactile learning—going through the experience of writing or typing the information, is just as valuable as auditory—listening or speaking it), building the electronic cards may hold just as much import in the learning process.

The last step is getting the file (probably named "flash-cards .htm" or something like that) onto your portable device. Since the project was first devised for the Sony PSP, let's look at that:

ELECTRONIC FLASH CARDS ON A SONY PSP

1. Use a USB cable to mount the PSP as a drive on your computer (may require a trip to the devices settings to allow it).

2. Create a folder in the root directory with a short, easy name (like "local"), and drop your flash card file in there.

3. Unmount the PSP.

4. Go to the PSP's Web browser and manually open the following URL: file:/local/flash_cards.htm. Please notice there is only one forward slash in the URL. For the savvy Web-heads, no, this is not a typo. It is rather a quirk of the Sony PSP's root access that took some deep Googling to uncover.

5. Having accessed the file, you should be up and running with the flash cards. I suggest you use the PSP's bookmarking ability to save the URL so you don't have to reenter it manually each time. I will also point out that knowing all this about root access of the PSP opens up the ability to save and access music, movie, and other files locally on the system. However, hacking your PSP is not what this book is about, so I'll leave it to you to explore some other time.

ELECTRONIC FLASH CARDS ON AN I-PHONE OR I-POD TOUCH

Devices from Apple are notoriously less accessible than their counterparts from other manufacturers, for both good and bad reasons. For our purposes, though, it doesn't matter. Indeed, the process is easier than the PSP but slightly more expensive, because you'll need to pick up a helper application. There are a variety of file storage and viewing apps available on the iTunes App Store for a variety of prices, but the two with which this project has been tested are Files and Pogoplug. Files costs $4.99 but is a great utility for accessing the storage available on your iPhone, and it allows you to upload files from your desktop/laptop computer to the phone for viewing. Pogoplug is free, but it requires you to own the $99 Pogoplug device for turning an external hard drive into network and Internet-accessible storage. Both these apps will let you store the flash card file on your device and run it. As more apps are tested, they will be listed on the www.geekdadbook.com Web site.

ELECTRONIC FLASH CARDS ON OTHER PORTABLE DEVICES

Realistically, any portable device with a browser and local file access should be able to use this tool. This includes all manner of smartphones as well as (rather obviously) netbook and laptop computers.

ELECTRONIC FLASH CARDS HOSTED REMOTELY

The alternative to actually carrying the file with you is, since the portable flash cards tool is in its essence a Web page, to host it on a personal Web site or other remote-access file storage and call it up whenever you have Internet access.

The portable flash cards tool is intended to be an open-source project, and you are welcome to take it, improve it, and adapt it for your needs.

Wi-Fi Signal Booster

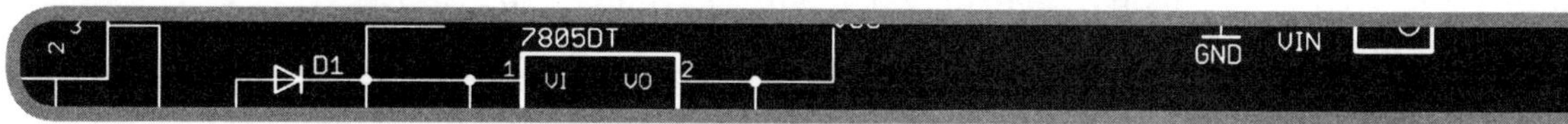

Most modern households, especially those with geeky parents and children who spend copious amounts of time connected to the world via the Internet, have wireless LAN setups with a router so the whole family can enjoy a wireless connection from anywhere in and around the house.

At least, that's the idea.

The reality is often something less functional. If a router is set up in the den/office, and Mom wants to connect and check her stock portfolio while sipping a mai tai in the sunroom, her connection may be fuzzy. Or if one of the kids wants to do research for his report on quantum vibration from the tree house out in the backyard, the signal may not be able to keep up with him. This is because the strength of the radio signal from a Wi-Fi base station drops off in steps over various distances (assuming nothing reflective gets in the way), and each step usually also degrades the available bandwidth. You need to maintain a strong signal to have a strong, higher-bandwidth connection with the base station.

It is possible to use repeaters—additional routers placed elsewhere in the house to repeat and amplify the signal—if you have a large house to cover, but that means buying extra hardware to place in additional locations, which could be a waste of money if you need the extra-strength connection only occasionally, or in a variety of locations. What if you could make yourself something with simple materials to boost the signal to your laptop and that you could move around the house with you as needed, and for less money than an extra wireless router? You can!

PROJECT	WI-FI SIGNAL BOOSTER
CONCEPT	Make a dish-type radio collector to help boost the signal strength from your household Wi-Fi routers to a USB wireless card on your laptop.
COST	$$$
DIFFICULTY	⚙⚙⚙
DURATION	☼☼–☼☼☼
REUSABILITY	🌐
TOOLS & MATERIALS	Metal mixing bowl, USB Wi-Fi adapter, USB extender cable, camera tripod, rubber bands, 1/4-inch nut, drill with various bits

If it isn't yours, there is likely at least one house in your neighborhood that gets satellite television. While the modern dishes are rather more compact than the van-size versions of the past, they all use the same concept: A concave reflector gathers and focuses electromagnetic waves of a particular range of frequencies into a receiver for translation into useable data. They point at a satellite out in orbit that is blaring its signal toward the side of the Earth facing it, and the dishes collect those signals and reflect them to a middle point for greater strength of reception. Our receiver will do pretty much the same thing with the radio signals given off by your Wi-Fi router.

There are three key elements to this project—the bowl, the USB Wi-Fi adaptor, and the USB extender cable. Everything else is about how to make the setup useful and mobile, and you may want to play with other ideas.

The bowl doesn't actually have to be a mixing bowl, but those are easily available in a variety of sizes, and while I haven't done extensive testing, it would make sense that the larger the bowl you choose, the better the results in signal boost. You can also use a pan or skillet lid, but whatever you choose should be metallic and concave.

STEP 1: To start, drill two holes in the bowl, one at the exact center (where, if you use a pan lid, there may already be a starter hole from the handle/knob you'll have to remove), for the Wi-Fi adapter, and one partway down the side for the tripod mount. Carefully measure the adapter, and it will take a little drilling and filing skill to get the center hole right, because the cross-section profile of the adapter will probably be more of a rectangular ellipse than a circle. You should draw the profile on the bottom with a black marker, then drill two to three holes to clear most of it, and use a metal file to finish it off. If you're going to use a camera tripod as your base, the second hole you drill should be ¼ inch, located about a third of the way down the side of the bowl or lid. Try to make sure it is located on a line that is perpendicular to the wide axis of the adapter hole (math geeks should understand that sentence; if you need help, go find a math geek and have him or her explain it to you).

STEP 2: To set the Wi-Fi adapter in the bowl, start by taking a standard rubber band and loop it repeatedly around the adapter about a third of the way from the receiving end (not the end with the USB jack), so that it becomes like a gasket. Depending on the rubber bands you have and how clean the hole you drilled is, you may want to use more than one band. Slip the adapter, jack end first, through the hole from the "inside" of the bowl, until the band stops it. This should leave the receiver end at the center of focus for your "dish."

STEP 3: Hold this construct together, and on the other side, instruct your child to take another rubber band and repeat the looping process until you have a second gasket, and roll it tight against the bottom of the bowl so the adapter is securely set. If you don't mind a more permanent solution, you could also use a hot-glue gun to set the adapter in place, though that won't let you take it back out to show off before/after readings to your incredulous (and slightly worried) friends and family.

STEP 4: To mount your dish to the tripod, simply set it on the camera head so that the threaded screw that would normally connect to the bottom of a camera pokes through the ¼-inch hole you drilled. Use the ¼-inch nut to hold the bowl securely on, tightening with a wrench if needed. Try to use a tripod with a metallic screw rather than a plastic screw, to avoid stripping the threads when tightening. Now you should have a very versatile means for aiming your dish.

STEP 5: When you're ready to test, there are a number of programs available on the Net to see detailed information on signal strength and available bandwidth. Take your dish to the place with a low signal, and set it up so that it is aimed roughly at where your wireless router is located as if you had line-of-sight through whatever walls or floors might be between. Connect the adapter to your computer with the extension cable, and start surfing with renewed vigor!

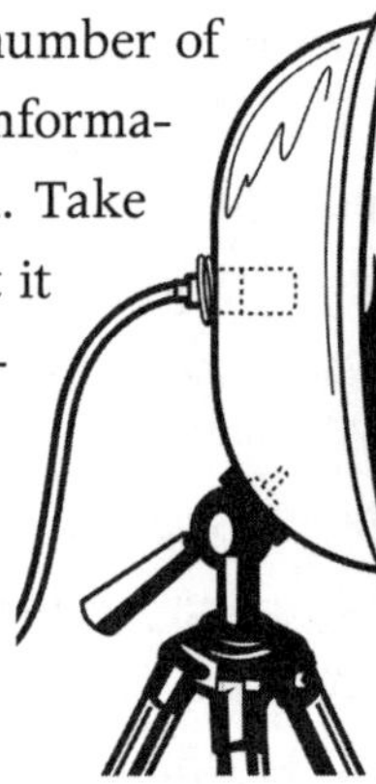

Cool LEGO Lighting from Repurposed Parts

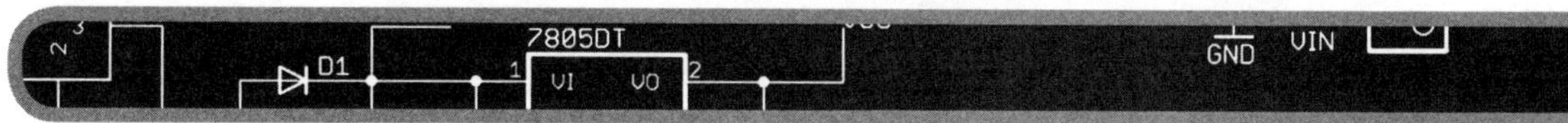

Reuse is an important concept these days. It's not just about recycling; repurposing existing items like furniture or building materials is a great way to minimize both waste and your carbon footprint.

And with the digital revolution in media, what's one thing we probably have lying around the house gathering dust? Yeah, CDs and DVDs. Hopefully, by now you've ripped all your old CDs into digital format for easy portability, and if you're a proper technology geek, you may have all your movies and music stored on a home server for access all over your local network, and even anywhere you've got an Internet connection. So why are you holding on to all those discs? Sure, you could try to trade them in at your local used-music store for some kind of credit, but why not build something with them instead? And why not make it a fun project you can share with your kids?

PROJECT	COOL LEGO LIGHTING FROM REPURPOSED PARTS
CONCEPT	Use LEGO, old media discs, an Arduino board, and a BlinkM LED unit to make a programmable ambient lamp powered off a USB port.
COST	$$–$$$$
DIFFICULTY	⚙⚙
DURATION	☼–☼☼
REUSABILITY	⊕
TOOLS & MATERIALS	LEGO, CDs and/or DVDs, Arduino board, BlinkM unit (or similar), computer

This project is fun because it lets you play with LEGO bricks (always fun!). But beyond that and in addition to the lesson of the importance of repurposing objects, you can also dabble in some pretty cool open-source electronics with your kid while constructing this lamp.

One of the coolest open-source initiatives out there is putting practical programmable electronics into the hands of hobbyists. It's called Arduino.

From the Arduino Web site (www.arduino.cc):

> Arduino is an open-source electronics prototyping platform based on flexible, easy-to-use hardware and software. It's intended for artists, designers, hobbyists, and anyone interested in creating interactive objects or environments.
>
> Arduino can sense the environment by receiving input from a variety of sensors and can affect its surroundings by controlling lights, motors, and other actuators. The microcontroller on the board is programmed using the Arduino programming language (based on Wiring) and the Arduino development environment (based on Processing). Arduino projects can be stand-alone or they can communicate with software running on a computer (e.g. Flash, Processing, MaxMSP).

The boards can be built by hand or purchased preassembled; the software can be downloaded for free.

What this means in plain(er) language is that you can buy an Arduino board, hook it up to your home computer, download and run some free software, and actually program the chips on the board to do different things with modules you attach to the board. It will help you teach your child that all the chips and wires crammed into every piece of home electronics you own aren't really magic boxes, but instead they are simple devices that one could easily learn how to hack with the right tools.

Before we get started, let me say that you don't have to build this project with an Arduino board. What I mean is that the way I'm showing you to do this project is nowhere near the only way you could do it, and if you're not in a mood to start learning programmable electronics, you don't have to. You can take this basic concept and find some other form of bright LED to use as the illumination for the lamp. Obviously, you'll have to play with the dimensions of the base to make sure everything fits the way you need, but that's half the fun of these projects—working together to figure out how to do it. I still recommend using LEDs, since most other forms of light will also give off significant heat, which could be problematic with plastic LEGO as your lamp shell. Be smart and careful!

But I urge you to consider the Arduino. Think of it as a gateway drug for promising electronics junkies. Once you and your kid have worked through the instructions for setting up and programming the board and light, you'll have started down a road of learning and discovery that will demystify every other gadget you ever own, and encourage a sense of invention and ownership that most people never have.

BUILDING THE LIGHT

For this project, we're using the Arduino Duemilanove board, with a BlinkM Smart LED. The Arduino board can be found for around $30 and the BlinkM for under $15 many places all over the Web, though I got both of mine from the kind folks at www.makershed.com. Another great resource for these boards and many incredible basic electronics projects is www.adafruit.com.

I'm not going to go through the detailed instructions on setting these up and programming them, for the directions are available online. You should get the appropriate links when you receive the parts (and see Appendix A for some links as well). The short of it is as follows:

1. Download and load the Arduino software onto your computer (Most OSes available).
2. Download the BlinkM Arduino script and controller application.
3. Hook up the Arduino board via USB.
4. Open the Arduino software.
5. Load the BlinkM script and upload it to the board.
6. Quit the Arduino software and detach the USB.
7. Attach the BlinkM to the Arduino board.
8. Reattach the USB.
9. Start up the BlinkM controller application.

10. Program the LED colors and sequence.

11. Upload the program to the board.

12. Quit the BlinkM controller application. The board just takes power from the USB, but the program runs natively.

Now you have your light source, programmed however you want it, powered by USB. All we need is to build the lamp structure into which to place it.

BUILDING THE LAMP

Once again we turn to LEGO as our favorite geeky building material. The lamp has to have two key sections: the base, into which the electronics are set, and the disc area, where the old media will be stacked. Since the Arduino and attached BlinkM boards are much smaller, the controlling dimension of the entire build is the diameter of the discs, which is approximately 15 LEGO studs across. To allow for 2-by-2 vertical posts to hold the discs in place, I created my base 20 studs by 20 studs. You may need to start on a larger base plate, or a number of smaller base plates interlocked to get to the correct size, depending on the bricks you have handy. Obviously, this is the time for improvisation, because the plates and bricks you have available will determine what your lamp looks like.

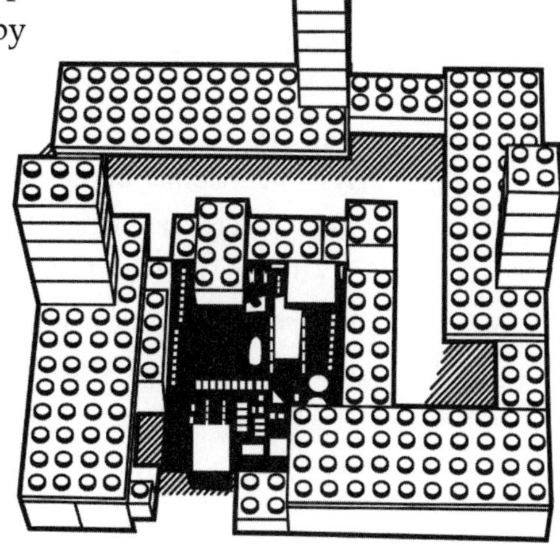

The base needs to be only 2 standard LEGO blocks tall. The challenge is to set the electronics board into the base

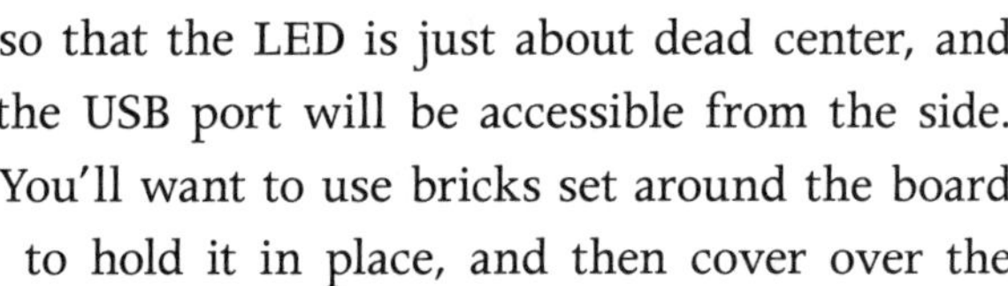

so that the LED is just about dead center, and the USB port will be accessible from the side. You'll want to use bricks set around the board to hold it in place, and then cover over the base structure with another layer of places that leave just the LED exposed.

At the midpoint of all four sides, you'll build 2-brick by 2-brick columns to contain the discs. You can go as high as you want, depending on the number of discs you have to use, but anything above about 6 bricks tall will make the light from the LED too diffuse to be effective.

Now stack your discs until they are just shy of the top of your columns. Put a 2-stud by 4-stud brick at the top of each column, with the extra length facing toward the center to hold the discs in place. Plug in your USB cable, and let the light shine!

ALTERNATIVE CONSTRUCTIONS

As I mentioned above, this isn't the only way to achieve the same idea. There are a number of alternative CD lamp designs on the Web, most of which don't use the electronics included here, opting rather for drilling out the center core of the discs to fit a compact fluorescent bulb, and making the base bigger to allow for the wiring to light it from a standard outlet.

An Even Cooler Idea!

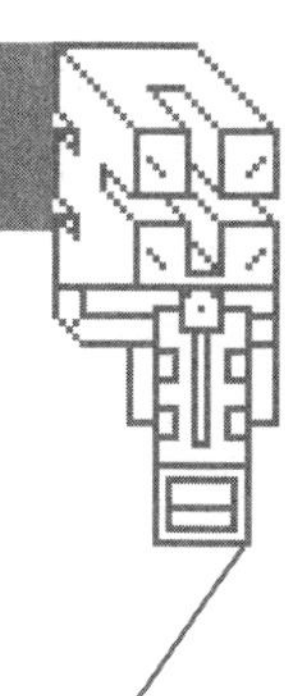

Another choice in the Arduino family is the BlinkM MaxM board. It uses much more powerful and larger LEDs, but it is programmable via the same software. It has the added feature of not requiring the Arduino board as a power source via the USB connector. Instead, the MaxM can take power from an external AC adaptor, allowing you to power your lamp from a standard outlet, or via batteries.

GEEKY POTPOURRI

Ice Cubes Fit for a Geek

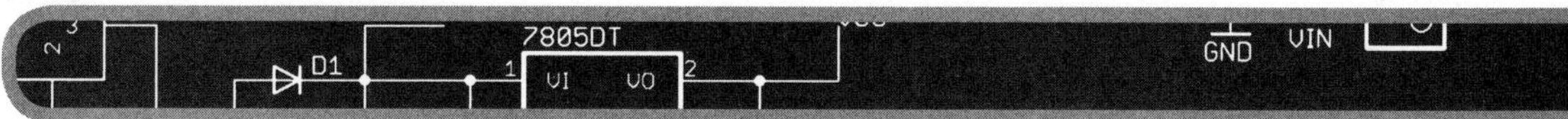

Ah, the simple LEGO brick! While the breadth of available LEGO parts is actually quite huge and varied (consider all the specialty pieces available for *Star Wars* and Indiana Jones sets, the Technics, Pirates, Power Miners, Mindstorms, and even Duplo sets, all with their own individual segments), the simple 2-stud by 4-stud brick is truly the iconic size. See one of those, and you know it's LEGO.

Heck, I'll bet there are lost bricks underneath some of the furniture in your house right now. You probably build sets with your kids, maybe even film stop-motion videos of the constructs (well, maybe that's just me). But have you ever wondered how else LEGO could be incorporated into your world? How about into the very beverages you drink?

PROJECT	ICE CUBES FIT FOR A GEEK
CONCEPT	Use silicon casting compound and LEGO to make LEGO-shaped ice cubes.
COST	$$–$$$
DIFFICULTY	
DURATION	
REUSABILITY	
TOOLS & MATERIALS	MoldRite 25 silicone casting mix, LEGO, petroleum jelly, X-Acto knife

[This project was developed by GeekDad writer Brian Little.]

First things first: Making your own LEGO ice cube trays is definitely not cheaper than buying them straight out from LEGO. But it is definitely more fun, and it is a great project to share with your kids.

MoldRite 25 is a two-part tin cure silicone rubber molding compound that's food-safe. You may need to do a little legwork to find it. It may be available at your local crafts store, or it may not. I ended up having to order it online and having it shipped, making it a more expensive project.

STEP 1: You can choose very simple blocks to model, or get a bit more creative. To keep it easy, use 2-by-2 and 2-by-4 (the most iconic) bricks. If you have a few other interesting shapes you want to try (being a geek is all about experimentation), think carefully about what will work well in creating a mold. Keep in mind that the shapes cannot be too complicated or have strange voids in them that will be difficult for the molding compound to seep into, or a challenge to get release from, when it is cured.

STEP 2: After you choose your bricks, wash them, dry them, and inspect them to make sure they are clean and unblemished.

STEP 3: I assembled three molds. The first one I created by building a 2-brick-high square wall, then laying single 4-by-2 bricks inside, spaced one stud apart. For the second, I used 2-by-2 bricks, and for the last, 1-by-1 bricks, so we could have a variety of ice cube sizes. Build your molds to accommodate slightly more compound than you actually plan to use. Filling the mold only partway up a brick results in a smoother edge.

STEP 4: Make sure the bricks are all pressed down and very tightly attached to each other to avoid any cracks for the molding compound to seep into. You can try to seal all the edges with the petroleum jelly, but leave as little residue as possible so the mold is not deformed (this may be a challenge—try using a Q-tip to apply).

STEP 5: There may be instructions on the molding compound suggesting you use a release compound, but I've found that with the LEGO, it isn't necessary. I also did not de-air the molding compound (another suggestion in most enclosed instructions), since I wasn't casting anything particularly detailed.

STEP 6: Fill the molds carefully. The compound is fairly thick and pours slowly, so be patient and cautious. You might try using a toothpick to distribute the mold evenly as you go, like cake batter. Try hard not to overfill so you get a cleaner base and no spillage.

Helpful Hints

If you use a kitchen measuring utensil to measure or mix your molding compound or catalyst, clean it up right away. I forgot to rinse the cup I measured the catalyst with, and the GeekSpouse was not impressed by what it took to clean it out.

Also remember to put down a piece of paper so you don't spill liquid rubber on the countertops.

STEP 7: Put all the molds aside to cure for 24 hours, but keep an eye on them, as they may be set enough to release in 12 hours (results may vary with different temperatures, relative humidity, and elevations). Be prepared for the end result to be, shall we say, not quite perfect. There may be a fair amount of flash (the technical term for the, um, gunk at the edges) that needs trimming off with the old X-Acto. However, you may be excited to see the resolution with which the bricks are detailed, right down to the tiny LEGO embossed on the top of every stud.

STEP 8: Now just fill them with water and stick them into the freezer. When freezing your ice, place the molds on a cookie sheet or other flat surface to help keep them even. Voilà! Your soda/ice tea/high ball will show to all around you just how much of a geek you really are!

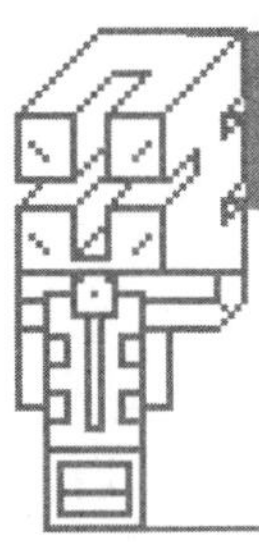

An Even Cooler Idea!

For more interesting LEGO brick ice cubes, try boiling some distilled water to get clear blocks, or dye the ice with food coloring (different-colored ice to match the variety of LEGO colors).

Exploding Drink Practical Joke

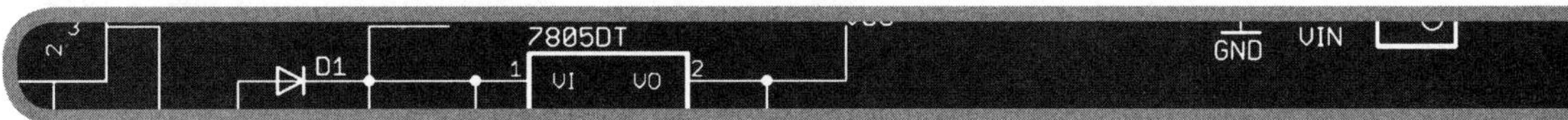

One of the finest things for a dad to share with his kid is the tradition of the practical joke. But since we are geeky parents, we won't debase ourselves with the tired jokes like the old flaming bag of dog poop on the doorstep. Rather, GeekDad practical jokes must include one very key ingredient to be worthy of our time and effort: science! Which is why this next project is for the geek in all of us.

PROJECT	EXPLODING DRINK PRACTICAL JOKE
CONCEPT	Make time-release "exploding" ice cubes.
COST	$
DIFFICULTY	⚙
DURATION	☼–☼☼
REUSABILITY	⊕⊕⊕⊕
TOOLS & MATERIALS	Ice cube tray, water, Mentos candy, soda (preferably Diet Coke)

Culled from the wilds of the Internet, this project/joke is based on the now (in)famous Diet Coke and Mentos reaction where some strange and wonderful interaction between the chalky hard candy and the carbonized beverage makes for an amazingly quick release of saturated CO_2 from the soda. But what the careful, scientific study performed by the MythBusters has taught us (episode #57) is that, while there is a chemical in Diet Coke that makes the reaction extra spectacular, we can get a similar, if slightly muted, effect from almost any carbonated soda.

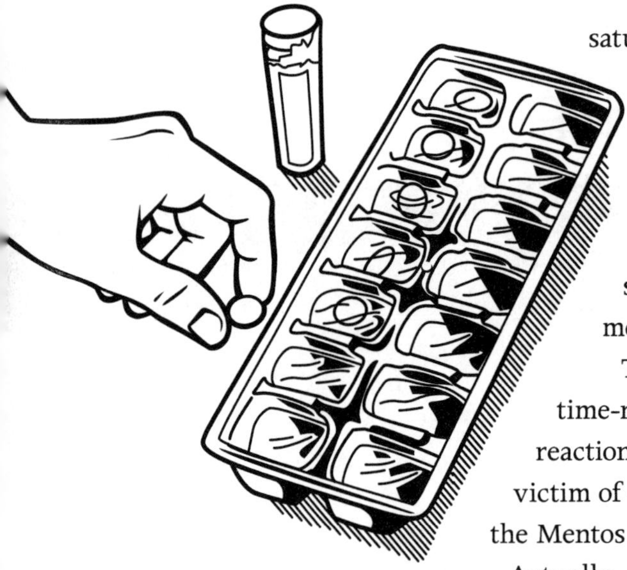

The concept is simple: We'll make a time-release system so that the Mentos/soda reaction comes as a complete surprise to the victim of our joke. To achieve this, we just freeze the Mentos into cubes of ice.

Actually, this is the trickiest part of the project (but not all that tricky). Fill your ice tray with water. Use only tap water, not distilled or filtered, because we want the ice cubes to be fairly opaque so as to hide the explosive contents. Put the tray into the freezer, and wait.

Give it about 10 minutes, and check to see if the freezing process has started. Almost always, ice cubes in trays freeze from the outside in. When you see a shell has formed on the top of your ice cubes, but they haven't frozen all the way through, you're ready. Use something like a butter knife to crack the top shell of the ice cube, and put one Mentos candy into the still-liquid interior of each cube. Top off the cubes with more water if needed, and return to the freezer to finish the extraction of heat energy. In about 30 minutes (depending on your freezer and how full it is), they should be ready.

Because of its volatile nature, this joke is probably best suited for the outdoors. Pick your victim, pour him the soda of choice (and hope he'll really want a Diet Coke, for the best fireworks possible), plop a special cube or two into the drink, and get the video camera ready (though far enough away to avoid splashing). YouTube and/or your child's therapist are waiting to see the hilarity that ensues!

Afterword

Pneumatic Wiffle Ball Cannon—Failure as a Project

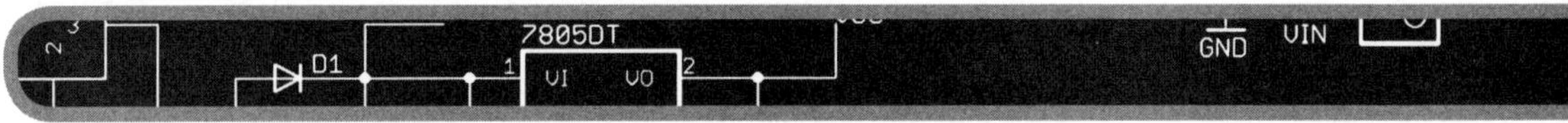

The best-laid plans of mice and GeekDads oft go awry, someone sort of said once. That's at least, if not more, true for the GeekDad who wants to design and build cool, geeky projects with his kid. And while working out a detailed design and layout for a project in advance is always a good idea, it doesn't always guarantee success. That's no reason not to try, though, and as we all know (anecdotally at least) we learn as much or more from failure as we do from success.

Which is why this project is included as an afterword. We failed at this project. It didn't work, and though I imagine someone could pull it together, we couldn't at the time, so decided to move on. However, we did walk away from it with something valuable.

PROJECT	PNEUMATIC WIFFLE BALL CANON—FAILURE AS A PROJECT
CONCEPT	Try to build a really cool project, fail, and learn from that failure.
COST	$–$$$$
DIFFICULTY	⚙ – ⚙⚙⚙⚙
DURATION	☼ – ☼☼☼☼
REUSABILITY	⊕⊕⊕⊕
TOOLS & MATERIALS	Pipe, fittings, screws, a saw, Wiffle balls, imagination, patience, and a sense of humor

I am not suggesting you try to build this project knowing that it will fail. It will fail if you follow what I did as described here. It was a fun idea that I thought I'd be able to develop into a cool, geeky, science-based construct with a wow factor for the kids. But it didn't work, for reasons those with a slightly better grasp of pneumatics will probably be able to easily describe. However, I'm going to use this project as a good excuse to discuss the value of failure as a learning tool, so please bear with me.

My older son plays Little League baseball, and it seemed to me that a cool idea would be to build some kind of neato baseball cannon for fun and practice. I had visions of hooking up some kind of PVC tube construct to my compressor and launching baseballs hundreds of feet into the air for some righteous outfield practice. I started doing some research on the Internet to find out who had done what before me, and learned that most of the work in this field has been done with more . . . explosive means toward launching the projectiles.

There were some versions using compressed air, however. The problem was that they all tended to use expensive valves and pressure chambers, which put the project out of the specs I was trying to keep all the GeekDad projects within. I needed to take a step back and reevaluate the project.

One day, watching my son at practice, I noticed that while they used regular baseballs for most of their drills, there was another kind they used in some cases. A lightbulb (LED) went on over my head: I'd build a Wiffle ball launcher!

Wiffle balls are so much lighter; I figured I could do something with manual compression. My boys and I went down to the local big-box home improvement store and dug through all the pipe and fitting they had. We returned home with 3-inch PVC, 4-inch and 6-inch drainpipe, a couple of rubber reducers, and, would you believe it, a toilet plunger. The concept was that I would build a compression chamber out of the larger pipe, which would reduce down to the 3-inch pipe. A Wiffle ball would be loaded into the 3-inch pipe and rest at its base on screws that had been drilled inward. The plunger fit just about perfectly into the 6-inch pipe and, the idea went, if you pushed the plunger upward quick enough, it would build enough pressure through the reducers to launch the Wiffle ball a significant height.

Yeah, well . . .

It was a complete failure. Being a civil engineer, I may have a basic grasp of certain areas of physics, but obviously my pneumatics needed updating. The boys and I were undaunted, though (well, maybe lightly daunted), because I had a fallback. I still had a compressor to work with.

We took apart most of the original build and instead devised something with a forechamber sealed mostly with duct tape. It had a small hole through which I could poke the end of the hand valve from the compressor's hose. With about 60 psi built up, it couldn't fail.

And again, so much for my memory of college physics. The Wiffle ball did actually make it out of the tube, but it flew in a very low arc, just a couple feet, and bounced on the ground. It was underwhelming, to say the least. I looked at my boys, they looked at me, and we all started laughing.

Time, and the cost of materials, made trying anything else prohibitive, so I asked them whether they were okay with calling it a fail. They were. They told me we'd tried our best, we'd improvised, explored the possibilities, but just couldn't make it work. And then they ran off to play with their buddy from across the street.

So I apologize to everyone reading this book who may have turned to this chapter, looking for a cool project to launch projectiles. It isn't here. I do firmly believe that it can be done, and probably in some way that's not too far off from what we tried. With another couple weeks of research on the Internet and fine-tuning a design, we probably could have made it work. But we couldn't, and it didn't, and that's okay. We tried, and we learned—both about what wouldn't work and about the necessary humility in any project where you're making it up from scratch. And I'm actually glad my boys and I could share that.

To make the long story just a little longer, my point is this: Be ready to fail. Indeed, when it comes, embrace failure as the learning experience it usually is. Don't get angry. Don't pound the workbench. Don't let your frustration get any bigger than a shake of the head and a rueful laugh. Show your geeklet that failure isn't an end but rather just another step toward ultimate success. Because patience and determination are vital components to the geek personality, and teaching that to your kids is as valuable a project as anything else in this book.

Appendix A

Resources and References by Chapter

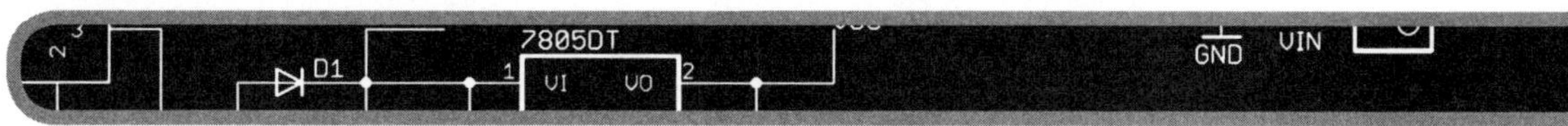

The instructions for the projects in this book are intended as a starting point for creativity and customization rather than a comprehensive guide. If you're interested in more information, and in sharing your variations on the projects in this book, the first place to go online is www.geekdadbook.com, where there are project pages and forums set up with an active, moderated community seeking to explore all the possibilities. Furthermore, the resources for many of the projects represented here are available online, and there are also alternative instructions on other Web sites. This appendix collects a list of links for each project to help you and your geeklets get as much as possible out of your creative experience.

I would like to extend special thanks to Phil Torrone and Marc de Vinc at www.makershed.com for supplying many of the electronics materials used in the projects. If you and your geeklets are interested in learning electronics as hobbyists, www.makershed.com and *Make* Magazine are THE places to start.

Another big thanks go to the folks at www.thinkgeek.com, who also supplied some materials for this book. www.thinkgeek.com is like the Toys-R-Us for geeks. Check it out!

RESOURCES AND REFERENCES BY CHAPTER

Make Your Own Cartoons

- Everything Photoshop, including a Web-based editor, is available at www.photoshop.com.
- Likewise, you can find the great Mac image editing program Pixelmator at www.pixelmator.com.
- And for free on all platforms, you can use GIMP, available at www.gimp.org.
- The program I used on the Mac to create the sample comic is Comic Life Magiq, easily found at www.plasq.com/comic-life-magiq.

The Coolest Homemade Coloring Books

- A great online resource for FREE print-it-yourself coloring pages is www.patternsforcolouring.com.

Create the Ultimate Board Game

- There is a dedicated section of the GeekDad book Web site for the board game at www.geekdadbook.com.
- A great site for more fun is www.boardgamegeek.com.

Electronic Origami

- You can buy the pen at www.frys.com/product/2931025.
- The LEDs are available at www.makershed.com or http://sparklelabs.com/index_store.php.
- You can find many more great origami folds at www.origami-club.com.

Cyborg Jack-o'-Lanterns and Other Holiday Decorations for Every Geeky Household

- The electronics mentioned are available at www.makershed.com.

Windup Toy Finger Painting

- For an extensive selection of windup toys, try www.tintoyarcade.com.
- Paints should be available at your local crafts store or at www.michaels.com.

Create a Superhero ABC Book

- As mentioned in the project, these are good sources for your heroes:

  The Superhero Dictionary (http://shdictionary.tripod.com)

  Comic Vine (www.comicvine.com)

  Marvel Universe: The Official Marvel Wiki (http://marvel.com)

- And if your geeklets really love superheroes and RPGs, try introducing them to Champions at www.herogames.com.

Model Building with Cake

- With the popularity of *Ace of Cakes* (www.foodnetwork.com/ace-of-cakes) and the success of Charm City Cakes (www.charmcitycakes.com), many local bakeries will build creative cakes to design.
- If you need help doing it yourself, talk to your local baker. We were lucky that ours (www.amiabakery.com) let us come in and use their airbrush to get the roof tiles and shrubbery right.

- Your local crafts store will probably have a baking section. We were able to find fondant and other supplies at the Michael's near us (www.michaels.com).

Pirate Cartography

- The original how-to for this project can be found at http://howto.wired.com/wiki/Make_A_Treasure_Map_From_A_Paper_Bag.

Parenting and Role-Playing Games

- Check out the forums at www.geekdadbook.com, where we hope to build a storehouse of ideas and lessons learned from parents and kids using this game.
- If you really want to get into RPGs with your kids, the home of D&D and a number of other great games is at Wizards of the Coast, www.wizards.com.

A Never-Ending Demolition Derby

- While you can get LEGO sets almost anywhere, the best place to get loose LEGO is at your local LEGO store or on the online store at www.LEGO.com.

See the World from the Sky

- Here's my original source for information on the lifting ability of helium: www.chem.hawaii.edu/uham/lift.html.
- Great source for extra-large latex balloons: http://anvente.com.
- When you're ready to go to the next level, try the BlimpDuino at http://diydrones.com.

Best Slip 'N Slide Ever

- You can find the plastic and hoses at your local hardware store, or online at places like www.hardwareandtools.com.
- The peel-and-stick Velcro can be found at Walmart with the picture hangers.
- Pool noodles are available almost everywhere that carries summer pool supplies.

Fireflies for Every Season

- If you don't have a local electronics store carrying simple LEDs, try www.frys.com/template/ecomponents.
- You can get cool multicolor LEDs, batteries, and other awesome electronics kits at www.adafruit.com.

Video Games That Come to Life

- For wooden hoops, check your local sewing/crafts store for inexpensive embroidery hoops.
- Online, go to a store like www.joann.com and search for "hoops." You should be able to find them as inexpensive as three hoops for a dollar.

Fly a Kite at Night

- There are a variety of kite stores online, including www.coastalkites.com, www.intothewind.com, and www.kittyhawk.com.
- I especially love the *Star Wars* kites available many places online, including www.thinkgeek.com/geektoys/games/a52a.

Build an Outdoor Movie Theater

- For more information on making your own outdoor theater, try http://backyardtheater.com.

The "Magic" Swing

- The phone books should be dropped off at your door.
- Rope and other fittings can be bought at your local hardware store or at www.hardwareandtools.com.
- The video of the original *MythBusters* episode "Phone Book Friction" can be found here: http://dsc.discovery.com/videos/mythbusters-phone-book-friction.html.

Smart Cuff Links

- You can find all the materials for this project at your local computer equipment store or try www.frys.com.

Light-up Duct Tape Wallet

- Much technical information was gleaned from the pages of www.ducttapeguys.com.
- If you'd rather buy your wallet, try www.thinkgeek.com or www.ducttapefashion.com.

Crocheted Dice Bag of Holding

- Try www.ravelry.com, a knitting and crochet community of online information and ideas.

The Science of Composting

- www.planetnatural.com is an excellent resource for composting materials and information.
- This project was inspired by the original GeekDad post www.wired.com/geekdad/2007/05/the_compost_bin.

Home Hydroponics

- www.planetnatural.com is also an excellent resource for hydroponics materials and information.
- Botanicalls kit: www.makershed.com/ProductDetails.asp?ProductCode=MKBT1. This is an awesome kit to allow your plants to tweet you when they need watering!
- Technical inspiration for this project came from www.hydroponics101.com and www.salviasource.org.

Build a Binary Calendar

- This project was inspired by the twelve-cent coin calendar described here: www.evilmadscientist.com/article.php/perpetualcalendar

Portable Electronic Flash Cards

- Here are two Web pages that were invaluable in figuring how to access local files on the PSP browser:

  http://asserttrue.com/articles/2007/02/02/how-to-run-local-flash-content-on-a-sony-psp. It explains the local URL (one slash, not two). It also tells you how to run Flash on the PSP, which is another possible way to implement the cards.

  www.brothercake.com/site/resources/reference/psp. This site gives great information on the capabilities and limitations of the PSP browser, specifically regarding CSS and Javascript.

Wi-Fi Signal Booster

- The Linksys WUSBF54G is one example of the USB Wi-Fi adapter mentioned in the project. You can find them all over the place.
- There are a number of variations of this project online. One of

my favorites is www.instructables.com/id/Wifi-Signal-Strainer-WokFi.

Cool LEGO Lighting from Repurposed Parts

- The Arduino board and BlinkM unit used for the project come from www.makershed.com.

Ice Cubes Fit for a Greek

- MoldRite is available from the maker's Web site, www.artmolds.com/, or on www.Amazon.com.

Exploding Drink Practical Joke

- The original DP & Mentos guys have a Web site at www.eepybird.com, where they sell specialty kits for getting really cool geyser effects like they do in their live shows.

Afterword: Pneumatic Wiffle Ball Cannon—Failure as a Project

- Someone succeeded at making a pneumatic baseball cannon here: www.wonderhowto.com/how-to-build-a-pneumatic-tennis-ball-cannon.
- *MythBusters*'s Adam Savage gave a great talk on failure at the 2009 Maker Faire, which you can watch at http://fora.tv/2009/05/30/MythBuster_Adam_Savages_Colossal_Failures. I was in the next building over at the time.

Appendix B

RPG Character Sheet

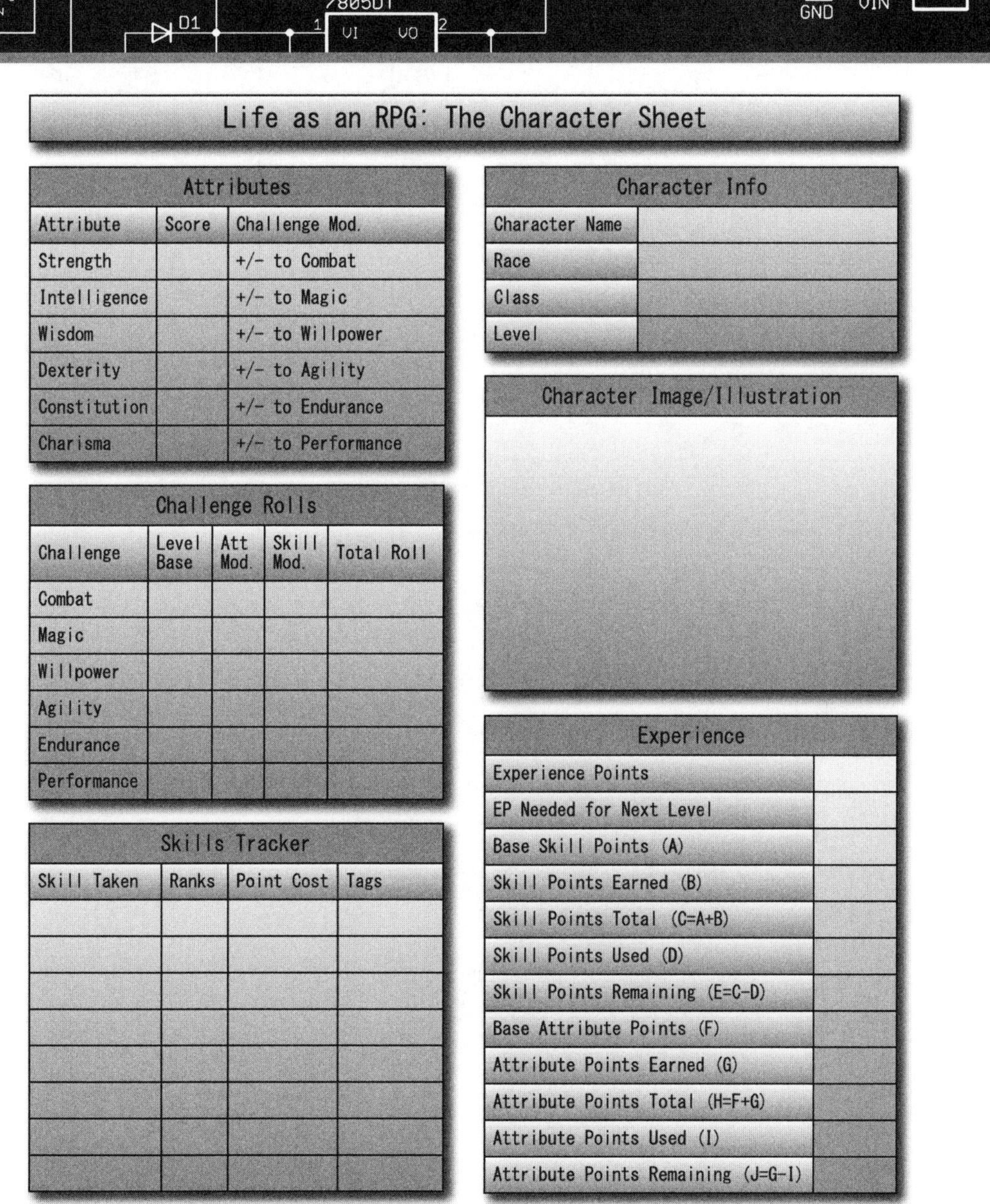

Life as an RPG: The Character Sheet

Attributes		
Attribute	Score	Challenge Mod.
Strength		+/- to Combat
Intelligence		+/- to Magic
Wisdom		+/- to Willpower
Dexterity		+/- to Agility
Constitution		+/- to Endurance
Charisma		+/- to Performance

Character Info	
Character Name	
Race	
Class	
Level	

Character Image/Illustration

Challenge Rolls				
Challenge	Level Base	Att Mod.	Skill Mod.	Total Roll
Combat				
Magic				
Willpower				
Agility				
Endurance				
Performance				

Experience	
Experience Points	
EP Needed for Next Level	
Base Skill Points (A)	
Skill Points Earned (B)	
Skill Points Total (C=A+B)	
Skill Points Used (D)	
Skill Points Remaining (E=C-D)	
Base Attribute Points (F)	
Attribute Points Earned (G)	
Attribute Points Total (H=F+G)	
Attribute Points Used (I)	
Attribute Points Remaining (J=G-I)	

Skills Tracker			
Skill Taken	Ranks	Point Cost	Tags

1 of 2

Life as an RPG: The Character Sheet

Challenge Journal

Challenge Description (Date)	Type	CR	Die Roll (A)	Chal. Roll (B)	Total (A+B)	Base Exp. (C)	Bonus (D)	Total Exp. (C+D)

2 of 2

Appendix C
Projects Listed by Rank

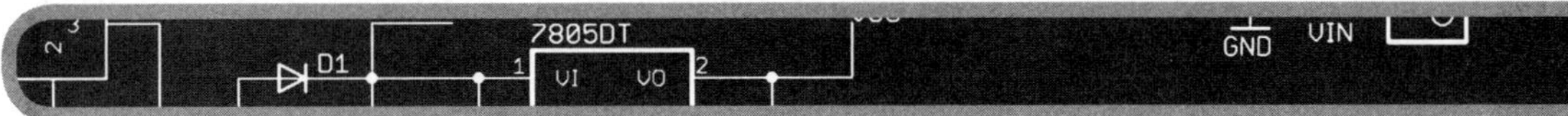

CHAPTER	COST	DIFFICULTY	DURATION
A Never-Ending Demolition Derby	$	⚙⚙	⏱⏱
Best Slip 'N Slide Ever	$$	⚙	⏱⏱
Build a Binary Calendar	$	⚙	⏱
Build an Outdoor Movie Theater	$$$$	⚙⚙⚙	⏱⏱
Cool LEGO Lighting from Repurposed Parts	$$	⚙⚙	⏱
Create a Superhero ABC Book	$	⚙	⏱
Create the Ultimate Board Game	$	⚙	⏱⏱⏱
Crocheted Dice Bag of Holding	$	⚙⚙⚙	⏱⏱⏱
Cyborg Jack-o'-Lanterns and Other Holiday Decorations for Every Geeky Household	$	⚙⚙	⏱
Electronic Origami	$	⚙⚙	⏱⏱
Exploding Drink Practical Joke	$	⚙	⏱
Fireflies for Every Season	$$	⚙	⏱
Fly a Kite at Night	$	⚙⚙	⏱
Ice Cubes Fit for a Greek	$$	⚙⚙	⏱⏱⏱⏱
Home Hydroponics	$	⚙⚙	⏱

CHAPTER	COST	DIFFICULTY	DURATION
Light-up Duct Tape Wallet	$	⚙⚙	⏱
Make Your Own Cartoons	$$	⚙⚙	⏱⏱
Model Building with Cake	$	⚙	⏱
Parenting and Role-Playing Games	$	⚙⚙	⏱⏱
Pirate Cartography	$	⚙	⏱
Portable Electronic Flash Cards	$	⚙⚙	⏱⏱⏱⏱
See the World from the Sky	$$	⚙	⏱
Smart Cuff Links	$	⚙⚙	⏱⏱
The "Magic" Swing	$$	⚙	⏱⏱
The Coolest Homemade Coloring Books	$	⚙⚙	⏱⏱
The Science of Composting	$$	⚙	⏱
Video Games That Come to Life	$	⚙	⏱
Wi-Fi Signal Booster	$$$	⚙⚙⚙	⏱⏱
Windup Toy Finger Painting	$	⚙	⏱